Donald T. J. Hurle

Crystal Pulling from the Melt

With 92 Figures

Springer-Verlag
Berlin Heidelberg New York
London Paris Tokyo
Hong Kong Barcelona Budapest

Dr. Donald T.J. Hurle

H.H. Wills Physics Laboratory
University of Bristol, Tyndall Avenue
Bristol BS8 1TL, England

ISBN-13:978-3-642-78210-7   e-ISBN-13:978-3-642-78208-4
DOI: 10.1007/978-3-642-78208-4

Library of Congress Cataloging-in-Publication Data
Hurle, D.T.J.
    Crystal pulling from the melt / Donald T.J. Hurle.
        p.   cm.
    Includes bibliographical references and index.
    ISBN-13:978-3-642-78210-7

    1. Crystal growth. 2. Crystal growth ·· Mathematical models.
    I. Title.

© Springer-Verlag Berlin Heidelberg 1993
Softcover reprint of the hardcover 1st edition 1993

The use of general descriptive names, registered names, trademarks, etc. in this publication does not imply, even in the absence of a specific statement, that such names are exempt from the relevant protective laws and regulations and therefore free for general use.

Typesetting: Macmillan India Ltd., Bangalore, 25 India;

51/3020-5 4 3 2 1 0 - Printed on acid-free paper.

Preface

Interest in crystals, and in particular in their form, dates back a long way but crystal growth as an industry belongs to the second half of the 20th century. Crystal pulling, the subject of this monograph, has its origin in a paper by Czochralski in 1918 but became exploited only with the arrival of the semi-conductor industry in the post Second World War period.

Initially crystal pulling equipment was designed purely empirically but as the industry has grown in importance and the size of crystals increased, consideration has been given to optimising the process. This has led to basic research into the fundamentals of the process together with a large number of computer simulations of elements of the process-particularly those concerned with heat and mass transport.

The purpose of this book is to attempt to strengthen the interface between the practical crystal grower seeking to improve or scale up his process and the applied mathematician who seeks to help him by analytical or computer modelling. The former individual is faced with the daunting task of understanding the fundamentals of convective flow and heat and mass transfer and the couplings between them if he is able to usefully interpret the results of computer simulations. The latter must acquire a sound understanding of what is relevant and necessary physics and chemistry to form the basis of his model if it is to have any value and not simply mislead the experimentalist. He must also acquire an understanding of the subtle compromises which the crystal grower has to make in order to optimise his process.

This book is not a guide to the technology and practice of crystal growth; these aspects are treated only at a level appropriate to understanding required by the modeller. Equally it is not a treatise on techniques of computer simulation such as computational fluid dynamics. Its focus is on the physics, chemistry and metallurgy of the process which is the point of contact between the crystal grower and the modeller. If it succeeds in convincing the grower of the value of modelling and enhances the veracity of that modelling it will have achieved its purpose. It should additionally be of value to undergraduate and graduate students taking crystal growth courses within various multidisciplinary study programmes.

I would like to acknowledge my gratitude to the many crystal growers and applied mathematicians who have helped me to an understanding of the practice and the theory of the crystal pulling process.

Finally I thank my wife Pamela for her encouragement and support and my daughter Katherine who kindly word-processed the manuscript.

D.T.J. Hurle

June 1993

Contents

1 Introduction and Historical Perspective

The elegant, but deceptively complex, technique of pulling single crystals from a melt by first dipping a seed crystal into that melt has been developed over the last forty years to become the dominant technique for the production of bulk single crystals of a wide range of materials for the electronics and electro-optic industries.

At the time of the famous Faraday Society meeting on crystal growth in 1949 it was virtually unknown. A review of melt growth techniques by Zerfoss et al. [1] at that conference included a description of the Kyropoulus̃ technique in which a seed crystal is dipped into a (fluxed) melt contained in a crucible which is then slowly cooled but the concept of steadily withdrawing that crystal from the melt was not considered.

However, just at that very time, Teal and Little at Bell Laboratories in Murray Hill, New Jersey, had just started their efforts to pull single crystals of germanium from the melt [2]. (Teal [3] writes that they made their first germanium single crystal in December 1948). The pioneering work of Czochralski [4], whose name is now associated with the technique, did not become widely known until the publication of Buckley's book "Crystal Growth" [5] in 1951. (Incidentally that book appears to be the first major compilation on the subject of synthetic crystal growth to have been published). It was not just that Czochralski's work was unappreciated; the whole subject of crystal growth from the melt lacked any scientific basis. Referring back again to the Proceedings of the 1949 Faraday Society meeting, one cannot find a single article, or a significant portion of an article, devoted to an understanding of the fundamentals of melt growth. In that era it was the external form of crystals which attracted interest and so it was the development of growth forms in unrestrained growth into supersaturated media which formed the focus for theoretical study.

Buckley reproduced a diagram of Czochralski's apparatus (Fig. 1.1) and devoted three pages of text to his work and to variants of the method due to von Gomperez [7] and to Mark, Polanyi and Schmid [8]. Buckley himself might not have known of this body of work had not Professor Polanyi come to work in the Chemistry Department of Buckley's own university (University of Manchester England) – and given him the surplus stock of his reprints (see Buckley's preface to his book) which no doubt included the papers referenced above. Czochralski's purpose was not to obtain bulk single crystals for some other experimental study but rather to measure the speed of crystallisation of a number of low melting point metals (tin, lead and zinc). He pulled single crystal wires of only 0.2, 0.5 and 1 mm diameter.

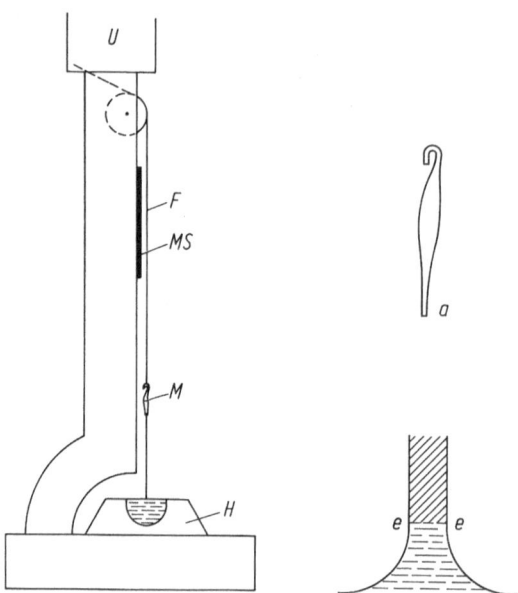

Fig. 1.1. Czochralski's original apparatus. The melt was contained in a charcoal crucible (*H*). Seeding was achieved by dipping a hooked capillary (*M*) into the melt. This was drawn up on a silk thread (*F*) by a clockwork motor (*U*), the rate of drawing of the metal wire being measured using a scale (*MS*). On the right hand side are shown the capillary seeding device (*a*) and the observed profile of the meniscus supported by the growing crystal (*e-e*). (Czochralski [4])

Credit for the development of the technique and its subsequent exploitation by the electronics industry belongs rather to Teal and Little. The history of their early work has been documented in an article by Gordon Teal [3]. He recalls that interest in obtaining single crystal germanium samples developed only after Haynes had shown that such material had minority carrier lifetimes some twenty to one hundred times greater than those of polycrystalline material. He reports that prior to that Shockley had been opposed to work on single crystals of germanium and had publicly stated on several occasions that he thought that transistor science could be elicited from small specimens cut from polycrystalline masses of material. The path of the innovative materials scientist has never been smooth!

The growth by pulling, of germanium single crystals and their characterisation is fully described in the classic series of books entitled "Transistor Technology" published by Bell Laboratories [9] in 1958 which provided the introduction to many of that generation into the world of semi-conductor electronics. Teal and Little may well not have been aware of Czochralski's work when they conceived the idea of pulling germanium from a melt. Czochralski must therefore count himself somewhat lucky to have been accorded the honour of having his name associated with a technique which has become part of the core technology of the world's most dynamic industry.

A diagram of the original Teal-Little crystal growing equipment is shown in Fig. 1.2. This was designed to be moveable since, being unable to get a separate laboratory in which to operate it, they were given permission to set the puller in the middle of a large metallurgical workshop to use it after normal working hours as long as they rolled the equipment into a closet when they had finished

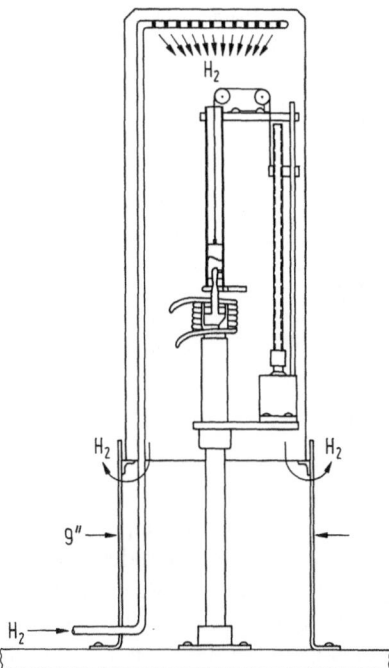

Fig. 1.2. Teal and Little's original equipment. The growth chamber consisted of a 30 inch high bell jar. Hydrogen was introduced at the top and flowed out at the bottom into the ventilating system of the room! The melt was contained in a graphite crucible which was RF-induction heated. Crystal lift was provided by a clock mechanism. (Teal [3])

each night so that it would not be in the way of the metallurgists during the daytime! This involved rolling the equipment and its associated high frequency heater some 25 to 30 feet into a corner of the shop.

The significant developments made by Teal and Little from Czochralski's original work were

a) the use of a "seed" crystal to define the crystal orientation
b) the control of the crystal's shape (i.e. increasing its diameter from that of the seed) by use of a programmed temperature control of the melt.
c) the control of dopant distribution by rotation of the seed crystal and by modulation of the pulling speed.

The next major application of the technique, after its deployment for germanium and silicon, followed from the discovery of the solid state laser. Pioneering work by Nassau and Broyer [10] and by van Uitert [11] on calcium tungstate followed by others [12] showed that the technique was capable of yielding crystals of higher optical perfection than the hitherto used, Verneuil technique. Pulling has become the standard method for the production of large crystals of a wide range of congruent and near-congruent refractory oxides. The topic has been reviewed by Brandle [6].

A variant of the pulling technique – known as liquid encapsulation – has revolutionised the production of III/V crystals and opened up major markets for some of these materials. The liquid encapsulation Czochralski (LEC) technique has become the dominant method for the production of single crystals of III/V

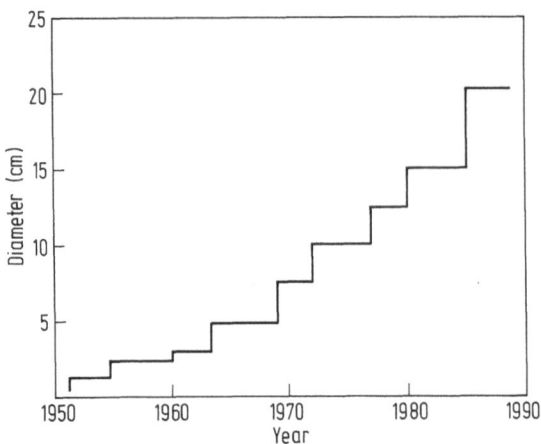

Fig. 1.3. Diameter of commercial silicon crystals vs year. (Siltec data)

compounds [27]. Its most important embodiment was the pressure puller [14], which enabled LEC growth to be applied to the Group III phosphides making available, for the first time, large high quality single crystals of these materials. Subsequently this same technique was applied to gallium arsenide to permit direct synthesis of the material from the elements in the puller. This has had a profound impact on the preparation of semi-insulating substrates for gallium arsenide integrated circuits [13].

Thus, in a relatively few years, the pulling technique arrived on the scene and swept across the whole gamut of electronic materials. The relentless progress is shown in Fig. 1.3 which charts increases in crystal diameter of commercial silicon.

Major advantages of the pulling technique include:

a) The crystal is unconstrained as it cools and a high structural perfection can be obtained. For example large single crystals of silicon containing no extended dislocations within their volume are routinely produced for device fabrication.
b) Crystal rotation ensures a uniform distribution of deliberately added solute.
c) The crystal can be observed throughout all stages of its growth.

2 Elements of the Process

2.1 Introduction

At its most basic (Fig. 2.1) the technique consists of a crucible which contains the charge material to be crystallised and a heater arrangement (shown as RF

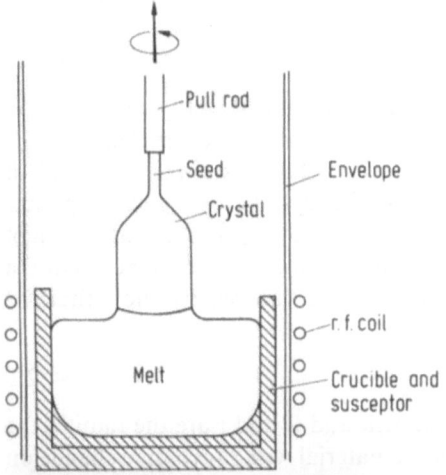

Fig. 2.1. Elements of the pulling process; configured such as might be used to grow germanium, silicon or low- melting point oxide crystals

Fig. 2.2. Research-scale silicon single crystal being pulled from the melt. The melt is contained in a silica crucible and is viewed from an inclined port in the side of the growth chamber. The RF heating coils which surround a graphite susceptor can be clearly seen

induction heating in the figure but various other forms are also in common use) to heat the crucible and charge to above the melting point of the latter. A pull rod with a chuck containing a seed crystal at its lower end is positioned axially above the crucible. This seed crystal is dipped into the melt and the melt temperature adjusted until a meniscus is supported. The pull rod is then slowly rotated and lifted and, by careful adjustment of the heater power, a crystal of the desired diameter can be grown. The whole assembly is maintained in an envelope which permits control of the ambient gas and enables visual observation of the crystal to be made. Figure 2.2 shows a small silicon single crystal being pulled from a melt contained in a silica crucible heated by RF induction.

2.2 Heat Balance and Growth Rate

The normal growth rate of the crystal-melt interface is essentially controlled by the rate at which the pull rod is raised – the 'pull rate', here denoted by v_p. If conditions are arranged so as to grow a cylindrical crystal (i.e. a crystal of constant radius) then the normal growth rate of a crystal growing with a planar crystal-melt interface (v) will be equal to the pulling rate v_p plus the rate of fall of the melt level in the crucible. The latter depends on the relative diameters of the crucible (assumed cylindrical) and the crystal. One can readily show that the growth rate is given by:

$$v = v_p \rho_L R^2 / (\rho_L R^2 - \rho_s r^2)$$ \hfill 2.1

where $\rho_{S,L}$ are the densities of crystal and melt and R and r are the radii of the crucible and crystal respectively. Thus, for a material with no volume change on solidification (equal densities), growth in a crucible which is twice the crystal radius occurs at a rate which is 33% greater than the pulling speed.

When the crystal radius is changing with time then the height of the supported meniscus is also changing and this makes an additional contribution to the growth rate as explained in Sect. 8.3.

To achieve the condition of steady growth with constant radius requires that the net heat flow through the interface exactly balances the rate of evolution of latent heat due to the crystallisation process. Again assuming for the moment a planar interface, this condition can be expressed simply as:

$$k_S G_S - k_L G_L = Lv$$ \hfill 2.2

where $k_{S,L}$ are the thermal conductivities of crystal and melt, $G_{S,L}$ are the axial temperature gradients in the crystal and melt at the interface and L is the latent heat of crystallisation per unit volume.

To summarise; the growth rate of the crystal is predetermined by the operator by choice of pulling speed and relative diameters of crystal and crucible and the growth of a cylindrical crystal is obtained by careful adjustment of the heat flow conditions so as to maintain the heat flux balance at the interface. This latter point is explained in more detail in the following section.

2.3 The Meniscus

The behaviour of the meniscus supported by the growing crystal controls the dynamics of the pulling process. Thus the shape of the crystal is determined by the angle at which the meniscus contacts the crystal at the three-phase boundary. This in turn is determined by the meniscus height which is itself determined by the thermal distribution.

The position and shape of the crystal- melt interface will conform to that of the isotherm corresponding to the crystal freezing temperature (which in general will be very close to the equilibrium melting point of the material.) An increase (decrease) in meniscus height brought about by an increase (decrease) in the heat supplied to the melt will cause the diameter of the crystal to decrease (increase). There is a unique meniscus height, corresponding to a unique thermal input, for which the crystal will grow as a right circular cylinder of given radius r. This is shown schematically in Fig. 2.3 where the contacting angle (Θ_L^0) is shown as being non-zero. This is characteristic of the growth of all diamond-cubic and zinc-blende semiconductors and of some oxide materials. On the other hand, metals, such as copper, contact the crystal with a zero wetting angle. The magnitude of the unique meniscus height which corresponds to a cylindrical crystal depends on crystal radius as displayed in Fig. 2.4. A non-zero angle is characteristic of materials which do not completely wet their own solids [16] i.e. which have the property that:

$$\sigma_{SG} < \sigma_{SL} + \sigma_{LG} \qquad\qquad 2.3$$

where σ_{SG}, σ_{LG} and σ_{SL} are respectively the crystal-gas, melt-gas and crystal-melt interfacial free energies. The melt-gas interfacial energy is numerically equal to the surface tension of the pure liquid but a distinction between surface energies and surface tensions for solids has to be noted [15].

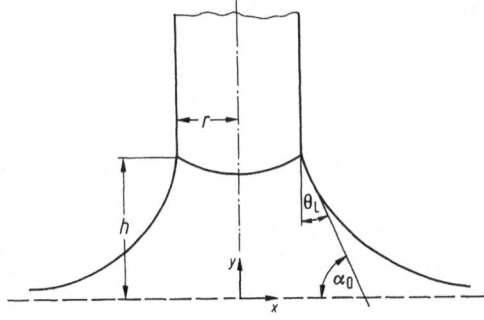

Fig. 2.3. Schematic representation of the meniscus supported by a growing crystal. Crystal radius r; meniscus height h; characteristic contacting angle $\Theta_L^0 = \pi/2 - \alpha_0$

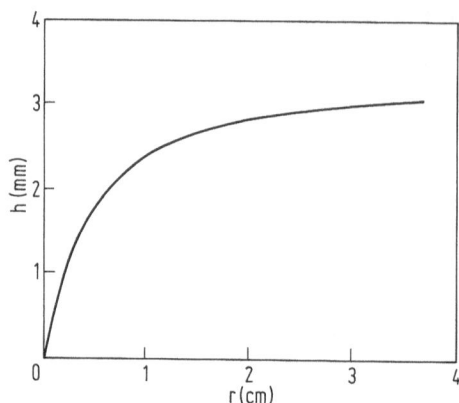

Fig. 2.4. Meniscus height versus crystal radius for the growth of a cylindrical crystal. (data appropriate to germanium)

Bardsley et al. [16] have shown that the Herring condition

$$\Sigma(\sigma_i t_i + \partial\sigma_i/\partial t_i) = 0 \qquad i = \text{SG, SL, LG} \qquad\qquad 2.4$$

expressing thermodynamic equilibrium along the three-phase line junction, can be used to give the contacting angles of the melt-gas and melt-crystal interface. (t_i is the surface normal). The resulting implicit expression for the contacting angle is

$$\cos\Theta_L^0 = (\bar\sigma_{SG}^2 + \sigma_{LG}^2 - \bar\sigma_{SL}^2)/2\bar\sigma_{SG}\sigma_{LG}$$
$$+ \{[\sigma_{LG}\sin\Theta_L^0/\bar\sigma_{SL} + \sin\Theta_I^0](\partial\sigma_{SG}/\partial\Theta)_0/\bar\sigma_{SG}$$
$$- [\sigma_{LG}\sin(\Theta_L^0 + \Theta_I^0)/\bar\sigma_{SG} - \sin\Theta_I^0]$$
$$\times (\partial\sigma_{SL}/\partial\Theta)_{\pi-\theta}/\bar\sigma_{SL}\}\bar\sigma_{SL}/2\sigma_{LG}$$

and for Θ_I:

$$\cos\Theta_I^0 = (\bar\sigma_{SG}^2 - \sigma_{LG}^2 + \bar\sigma_{SL}^2)/2\bar\sigma_{SG}\bar\sigma_{SL}$$
$$- \{[\sigma_{LG}\sin\Theta_L^0/\bar\sigma_{SL} + \sin\Theta_I^0](\partial\sigma_{SG}/\partial\Theta)_0/\bar\sigma_{SG}$$
$$- [\sigma_{LG}\sin(\Theta_L^0 + \Theta_I^0)/\bar\sigma_{SG} + \sin\Theta_I^0](\partial\sigma_{SL}/\partial\Theta)_{\pi-\theta}/\bar\sigma_{SL}\}/2 \qquad 2.5$$

where the bars indicate averaging round the periphery of the crystal. Because of the dependance of σ_{SG} and σ_{SL} on crystallographic orientation, the torque terms $(\partial\sigma_{SL}/\partial\Theta)_{\pi-\theta}$ and $(\partial\sigma_{SG}/\partial\Theta)_0$ will in general be non-zero. However, it was shown that they are unlikely to seriously effect the calculated value of Θ_L^0 so that neglect of them is justified.

The behaviour of materials which are completely wetted by their own melts is less well understood. For these materials it has been demonstrated experimentally that a thin liquid film will extend for a short distance up the side of the crystal. This phenomenon – which can be thought of as surface melting – has been observed on droplets of molten copper and a thermodynamic model has been derived [17] which predicts the thickness of the layer (in units of a mono-layer) to be:

$$n_0 = [q(\sigma_{SG} - \sigma_{SL} - \sigma_{LG})^{1/(q+1)}]/w\delta T \qquad 2.6$$

where w is the latent heat for the mono layer and q a constant with $q \approx 3$. Note that the layer thickness increases strongly as the undercooling δT approaches zero so that, on a sufficiently magnified scale, the three-phase boundary should appear as in Fig. 2.5.

The shape of the free liquid surface is given by solution of the Laplace-Young equation:

$$2z = -\beta\{z''/(1 + z'^2)^{3/2} + z'/x(1 + z'^2)^{1/2}\} \qquad 2.7$$

where z and x are respectively the vertical and radial coordinates. The primes signify differentiation with respect to the radial co-ordinate (x).

$$\beta = 2\sigma_{LG}/\delta\rho_L g \qquad 2.8$$

is known as the Laplace constant and $\delta\rho_L$ is the difference between the densities of the liquid and gas. g is the acceleration due to gravity. The liquid surface remote from the crystal lies in the plane $z = 0$. For the meniscus geometry this equation is not integrable analytically but Hurle [18] and Boucher and Jones [19] have derived approximate analytic solutions which are valuable for crystal growth simulation.

The Boucher and Jones expression is the most accurate but is an implicit relationship given by the equations:

$$Z^2(1 - X^{-1}) = 2(1 - \cos\alpha)$$

$$X = R + (1 + R^{-1})^{-1/2}\{\ln[\tan(\alpha_0/4)/\tan(\alpha/4)]$$

$$+ 2[\cos(\alpha_0/2) - \cos(\alpha/2)]\}$$

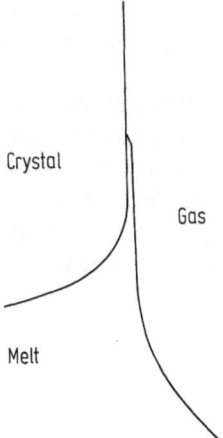

Crystal

Gas

Melt

Fig. 2.5. Form of the three-phase boundary for a material exhibiting surface wetting

where the co-ordinates x, z and the radius r are scaled by $(2/\beta)^{1/2}$ and denoted X, Z and R. The angle $\alpha = \pi/2 - \Theta_L$.

Hurle [18] has shown that further approximation yields the sufficiently accurate explicit relationship:

$$X = R + (2/A - H^2)^{1/2} - (2/A - Y^2)^{1/2} - \{\ln[Y(2^{1/2} + (2 - aH^2)^{1/2}]$$
$$- \ln[H(2^{1/2} + (2 - AY^2)^{1/2}]\}/(2A)^{1/2} \qquad 2.9$$

where $A = [1 + (\sin \alpha_0/RH]/2$
with the normalised meniscus height H given by:

$$H = \{2(1 - \cos \alpha_0) + [(\sin \alpha_0)/2R]^2\}^{1/2} \qquad 2.10$$

2.4 Crystal Rotation

The principal reasons for rotating the crystal are twofold. The first is the pragmatic one that doing so helps give the crystal the desired circular cross section. Without any rotation, the crystal would grow most rapidly in directions corresponding to minimum heat transport from the melt to the crystal so that, if the equipment did not have perfect thermal symmetry about a vertical axis coincident with the pulling axis, the crystal would wander off the pulling axis producing an object resembling the proverbial dog's hind leg.

The second, more subtle, reason for imposing crystal rotation concerns the distribution of dopant (deliberately added solute) into the crystal. Without any rotation the solute distribution in the melt would be subject to the vagaries of the natural convective flow in the melt (considered in detail in chapter 4). This flow is driven principally by the buoyancy produced by the temperature field so that, again, if this lacks perfect symmetry about the pulling axis, the flow pattern will also lack such symmetry. In Chap. 6 it is shown how the rotating crystal establishes a solute boundary layer ahead of it which has the property that its thickness is sensibly independent of radial position and thereby yields crystals having a radial uniformity of doping.

In addition to crystal rotation, crucible rotation is commonly employed for the commercial growth of semiconductor crystals but, curiously, not for oxide crystals. The reasons for this appear to be partly historical but the use of crucible rotation in silicon growth is also bound up with control of oxygen uptake into the crystal (see Chap. 7). Crucible rotation also influences the shape of the crystal-melt interface, a parameter which is of importance in controlling crystal perfection.

3 Techniques and Technology

3.1 Introduction

A detailed description of the process can only be made with respect to specific materials. The wide applicability of the technique – from materials like bismuth which melts at only 271 °C [76] up to refractory oxides melting at temperatures approaching 2500 °C [77] necessitates a diverse technology. There is no such thing as the universal crystal pulling machine. However, research pullers are usually modular in construction to permit fitment of different heaters, chambers etc. Commercial pullers, on the other hand, are designed for one specific material; indeed often for just one specific size and specification of single crystal of that material.

Thus a wide range of crucible materials are required for chemical compatability (i.e. inertness) with the melt. For example, quartz or graphite are commonly used for many metallic and semiconducting systems whilst precious metal crucibles are needed for many of the refractory oxides.

Crucible materials and melting temperatures frequently dictate the choice of heating system. Thus with high temperature materials using precious metal crucibles, RF induction heating is commonly employed. On the other hand, for reliability and low cost operation, commercial pullers for the growth of silicon and III/V compounds usually utilise "picket fence" high current, low voltage resistance heaters.

For highly reactive materials, such as the rare earth metals, specialist techniques have been developed. These include the use of a "cold crucible" which is RF heated and which, by inducing currents into the melt, produces a repulsion between melt and crucible wall which maintains the melt suspended, out of contact with the crucible [20], [21].

In the liquid encapsulation variant of the pulling technique, the melt is covered with a layer of a low melting point glass – commonly boric oxide – which, when an inert gas pressure exceeding the dissociation pressure of the melt is applied, prevents loss of the volatile component from the melt. Where this dissociation pressure is much above one atmosphere, then the puller chamber must be built to withstand this pressure. Viewing of the process is then provided by closed circuit T.V. through a quartz viewing port in this pressure chamber. On the other hand, where the process can be carried out at or below one atmosphere pressure, a simple silica envelope is commonly employed giving good visibility.

This book is not concerned principally with the technology of the pulling process but rather with the underlying science. Accordingly no attempt is made to provide detailed practical information on the design and operation of pulling systems. Consideration is confined to briefly reviewing the state of the art in the commercial production of the two most important semiconductors – silicon and gallium arsenide – and of high melting point oxide crystals.

3.2 Silicon

The charge weight and crystal diameter of silicon single crystals for commerical use have grown steadily (Fig. 1.3) since the first crystals were pulled in the early 1950 s. Current integrated circuit production is at 6 and 8 inches diameter from charges in excess of 50 kg. The crystals are now so massive that mechanical aids are required to remove them from the pullers! A more modest scale research puller is shown in Fig. 3.1.

Fig. 3.1. Research silicon puller capable of growing up to 7.5 cm diameter crystals from a 10 kg melt charge. The puller is shown fitted with a superconducting magnet producing an axial magnetic field. (Courtesy of K Barraclough)

The poly-silicon charge is placed in an ultra high purity synthetic quartz crucible which is supported in graphite of high purity which has been carefully pre-baked. In order to limit the height of the equipment the seed crystal is frequently attached to a chain or cable rather than to a pull rod, this chain or cable being wound up on a drum. Crystal rotation is achieved by rotating the complete drum assembly. A further control parameter is provided by rotating the crucible. The exact recipe for crystal and crucible rotation rates – and for changes made to them during the growth cycle – vary from manufacturer to manufacturer and are proprietary. The crucible is raised progressively as growth proceeds so as to maintain the position of the melt surface with respect to the heater.

Compared to other semiconductors, silicon enjoys two important advantages from a crystal growth point of view. These are a high thermal conductivity which permits rapid removal of the latent heat of solidification and a high critical resolved shear stress which means that quite high thermal gradients can be sustained by the cooling crystal without dislocation generation. (see Chap. 10) These two facts enable quite rapid growth speeds to be employed which is of considerable economic benefit. For example, 6 inch diameter crystals can be grown at a rate of 2.5 inches per hour. This is, for example, seven times faster than the speed that gallium arsenide crystals of only half that diameter can be grown.

The melting point of silicon is 1420 °C and this temperature is generated in the melt by the use of a graphite picket fence heater fed from a high current, low voltage A.C. or D.C. supply. Electrical power consumption for a 50 kg charge is in excess of 100 kilowatts.

Present day commercial pullers are highly automated with full computerised control. Most major producers of silicon have added their own in-house developed control systems proprietary to their own process. Much of this is concerned with subtle adjustments to the parameters as growth proceeds in order that the maximum amount of the grown ingot can be used to generate wafers which fall within steadily tightened specifications. As explained in Chap. 8, the Czochralski process can be inherently unstable to fluctuations in crystal diameter and it is necessary therefore to have servo-control systems to achieve the tight specification on diameter. Almost universally these utilise a photo- diode array which images a bright ring seen around the meniscus of the growing crystal [22] and which is, in fact, an image in the meniscus of the hot upper portion of the wall of the crucible. Annular movement of this bright ring heralds impending change in diameter (see Chap. 8) and this is used to control the diameter by modulating the crystal pulling speed. Changing the pulling speed changes the rate of liberation of latent heat which, in turn, controls the diameter of the crystal as explained in Chap. 2.

Process control is very much concerned with minimising the concentration of unwanted residual impurities such as the acceptor impurity boron and of controlling the concentration of the residual impurities oxygen and carbon and achieving uniformity of their distribution both on a macroscopic and on a microscopic scale. The oxygen concentration in particular is crucial in determining the properties of the material under heat treatment in a way which is central to the device fabrication technology [23]. Both the concentration and the micro-distribution of these impurities are intimately bound up with the convective flow in the molten silicon and, in recent years, much effort has been expended in understanding this. The topic is reviewed in detail in the next chapter and in Chap. 7. As part of a strategy to control this flow, the use of magnetic fields has been extensively investigated and the recently developed cusped magnetic field technology holds considerable promise for achieving improved control of the oxygen doping of the silicon ingot [24], [25]. As melt sizes increase progressively, then the strength of the natural convective flow also increases and with it the degree of turbulence in the melt. This turbulence generates temperature fluctuations which cause microscopic growth rate fluctuations which, in turn, lead to microscopic variations in dopant concentration known as solute striations.

Oxygen is introduced into the crystal as a result of erosion of the silica crucible by the molten silicon. Dissolved oxygen is transported by the convecting melt towards the crystal interface where it is incorporated into the crystal. However, most of this oxygen evaporates from the free-melt surface as silicon monoxide and this rate of loss can, in part, be controlled by controlling the ambient pressure in the gas. In fact, crystal pullers today are operated at a reduced pressure of a few kPa only. The formation and transport of carbon into the melt as carbon monoxide is also dependent on the ambient gas pressure.

3.3 Gallium Arsenide

Early attempts to apply the crystal pulling technique to gallium arsenide, which by this time had become so successfully and widely used in the silicon industry, were beset with difficulties in preventing dissociation of the gallium-arsenic melt. The method employed initially, known as a magnetically-coupled syringe puller [26], was an arrangement whereby the whole puller chamber was kept at a temperature above the condensation temperature of arsenic (around 630°C) so that a partial pressure of arsenic of around one or two atmospheres could be maintained within the growth chamber. A 50:50 gallium-arsenic melt held at the melting temperature of GaAs (1511 K) was maintained in equilibrium with that of arsenic vapour. Since there was no satisfactory engineering solution to the maintenance of a pressure seal operating at 630°C in an atmosphere of arsenic vapour, then the pulling and rotation mechanism had to be arranged remotely. This was achieved by having a sealed quartz growth chamber containing a precision bore section which acted as a bearing for a magnetically coupled crystal pulling rod. This rod contained a high Curie point alloy and this was coupled magnetically to an external magnetic field which provided rotation and lift. As the reader will readily imagine, this was not a technique which was easily adaptable as a commercial process since it required the careful assembly of large pieces of quartz-ware for each growth run.

The breakthrough came with the development of the liquid encapsulation Czochralski (LEC)-technique [27] which adopted the elegantly simple solution of covering the gallium-arsenic melt with a layer of an inert molten glass-boric oxide. This technique was pioneered by Mullin and co-workers at R.S.R.E., Malvern from a principle devised by Metz, Miller and Mazelsky [28] for the growth of lead chalcogenide crystals. The layer of boric oxide floats on the surface of the melt and the seed crystal is dipped through the boric oxide layer and equilibrated with the melt to achieve growth at the boric oxide-melt interface. (Fig. 3.2). Now provided that there exists an inert gas pressure in the chamber which exceeds the dissociation pressure of arsenic over molten gallium arsenide then the formation of gaseous arsenic is avoided and arsenic can escape from the melt only by diffusion through the viscous boric oxide in which it is virtually insoluble. The central advantage of this technique is that, unlike the magnetically-coupled syringe technique, the containing walls can be cooled and held at room temperature. This advantage is most apparent when growing the Group III phosphides where the dissociation pressures of phosphorus over the compounds are some tens of atmospheres. Pressure vessels operating at these pressures and at temperatures in excess of 600 °C are wholly impracticable and it was not until the advent of the LEC technique that good single crystals of these materials were grown [14], [29], [30]. A schematic representation of a pressurised system for the growth of GaP is shown in Fig. 3.3.

A crystal pulling system for the growth of gallium arsenide (for which the dissociation pressure of arsenic is around one to two atmospheres) can be a quite-conventional crystal puller, its operation being distinguished only by the layer of boric oxide on the melt surface. Synthetic quartz crucibles can be used but these tend to result in the uptake of some silicon into the crystal doping it 'n' type. Accordingly, more recent practice is to use crucibles fabricated in pyrolytic

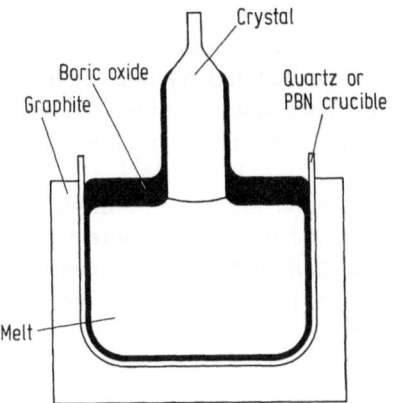

Fig. 3.2. Schematic representation of liquid encapsulation Czochralski (LEC) growth

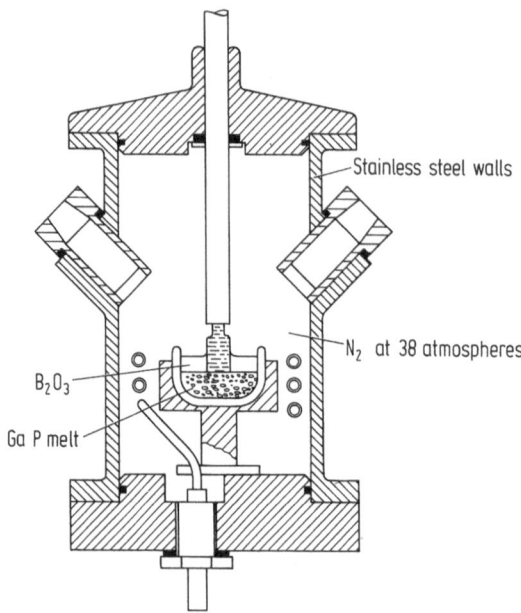

Fig. 3.3. Schematic representation of a pressure puller for LEC growth of gallium phosphide. (Mullin et al. [14])

Stainless steel walls

N_2 at 38 atmospheres

B_2O_3

Ga P melt

boron nitride. The quartz or boron nitride crucible is supported in graphite and either RF heating or a graphite resistance heater utilised. In commercial systems however, on grounds of servicability and cost, graphite resistance heaters are almost universally used.

The above constitutes what is known as the low pressure LEC technique. However, most of the world's commerical semi-insulating gallium arsenide for micro-wave and integrated circuit applications is currently produced by a high pressure LEC technique in which the growth chamber is a stainless steel pressure vessel capable of operating at pressures up to around 100–200 atmospheres (see [31] for a recent review). The merit of this approach is that it permits direct synthesis of the gallium-arsenic melt from the elements within the crystal puller. In the low pressure technique the gallium arsenide charge material was synthesised by distilling arsenic into a gallium melt and crystallising it in a separate apparatus. However this led to some silicon contamination of the melt again rendering it 'n' type. Semi-insulating gallium arsenide required for integrated circuits, could be made from this material only by doping it with chromium. The high diffusivity of chromium however made this an unsatisfactory process for some device fabrication stages and the high pressure process came into favour when it was demonstrated that undoped semi-insulating material could be grown from silicon-free gallium arsenide charges which had been directly synthesised in a pyrolytic boron nitride crucible in a high pressure puller [75, 78].

Subsequent to this, the low pressure technique has been extended to permit direct synthesis by distilling arsenic into the melt through a bubbler from a solid arsenic source contained in a separate chamber within the puller (see Fig. 3.4). However this technique has largely failed to win back the market from the high pressure process, the principal reason being that control of single crystallinity under low pressure conditions is exceedingly difficult. Added to this, the thermal stability of the semi-insulating material produced by this technique is sometimes less reliable than that of the high pressure process.

Commercial state-of-the-art machines are capable of accepting charges from 15 to 25 kg and growing material at up to 10 cm in diameter. The crucible is heated with a graphite "picket fence" resistance heater and optionally a second after-heater assembly can be included to reduce the thermal stresses in the cooling crystal. Diameter control is available based on a servo-control system using crystal weight measured by suspending the pull rod on the end of a load cell. The principles of this technology are described in detail in Chap. 8.

The growth procedure for undoped semi-insulating crystals is as follows. The pyrolytic boron nitride crucible is charged first with solid arsenic and then with rods of gallium and the charge is capped with a pre-formed disc of boric oxide which has been purified and then dried to a controlled water content typically in the range 200 to 2000 ppm. (Fig. 3.5). The puller is then pressurised with dry argon.

The crucible resistance heater is used to raise the crucible temperature first to around 450 °C at which point the boric oxide glass softens and flows over the charge of arsenic and the, now molten, gallium. The ambient argon pressure is

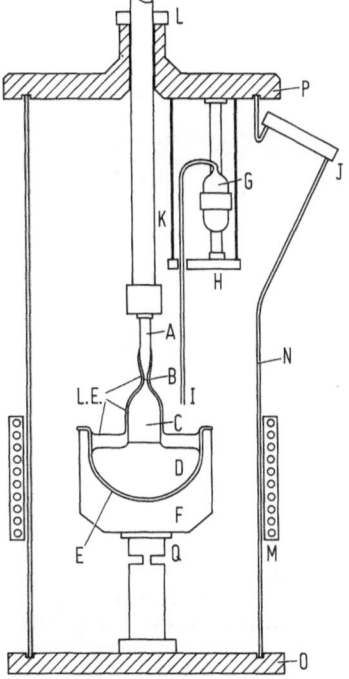

Fig. 3.4. LEC puller with arsenic reservoir for charge synthesis. (Mullin [29])

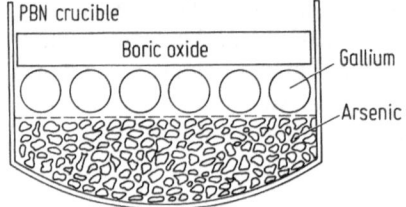

Fig. 3.5. Loading of the charge in a high pressure LEC puller for the growth of GaAs. [250]

then raised to around 60 atmospheres and the temperature of the crucible raised to around 800 °C at which point the gallium reacts exothermically with the arsenic raising the temperature further. Time and care must be taken in this synthesis stage to control the rate of heat evolution. The charge temperature is then progressively raised until the whole charge is molten at a temperature of 1238 °C. The chamber pressure can now be lowered to around 20 atmospheres for the growth phase. Some small loss of arsenic inevitably occurs during synthesis but this is within acceptable limits although more stringent specifications on material in the future may require more precise control of crystal stoichiometry.

A (100) oriented seed crystal is attached to the pull rod by a chuck which is slowly lowered through the boric oxide to the melt surface and the melt temperature carefully adjusted until a meniscus is supported by the seed crystal.

Fig. 3.6. Commercial high pressure LEC puller capable of growing 10 cm diameter GaAs from a 25 Kg charge. (Courtesy of Metals Research Semiconductors Ltd.)

A simple electrical circuit can be used to indicate when seed contact with the melt has been established [231]. Seed crystal and crucible are both rotated at rates in the range 0 to ± 20 rpm determined empirically by each manufacturer. The crystal is pulled from the melt at a rate which is usually less than 1 cm per hour. Crucible lift is employed to maintain the melt surface level with respect to the heater. Drainage of the boric oxide from the crystal as it emerges is slow so that the crystal can be essentially encapsulated in the molten glass. This reduces surface degradation of the crystal due to evaporative loss of arsenic. A commerical high pressure LEC puller for the growth of semi-insulating GaAs is shown in Fig. 3.6. Current commercially produced material is up to 100 mm in diameter (Fig. 3.7). Polished wafers from such material are shown in Fig. 3.8.

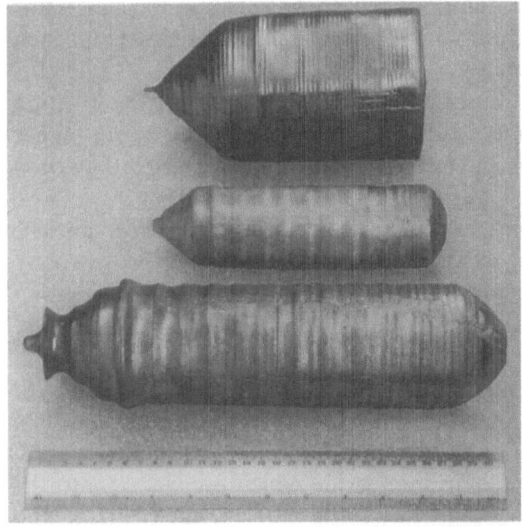

Fig. 3.7. Single crystals of GaAs 5-, 7.5- and 10 cm in diameter (courtesy of MCP Wafer Technology Ltd.)

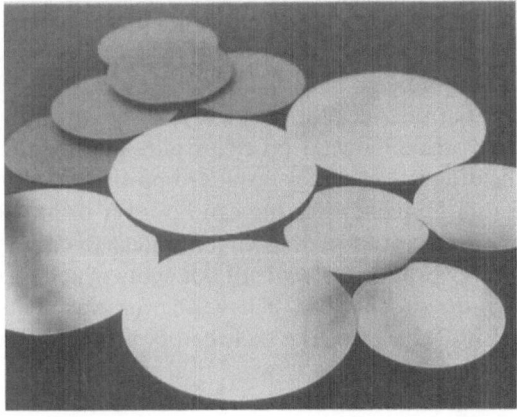

Fig. 3.8. Polished wafers of GaAs. (Courtesy of MCP Wafer Technology Ltd)

Axial or transverse magnetic fields can be applied to the melt during growth in order to control the melt turbulence and dopant segregation and this is discussed in detail in Chap. 7.

3.4 Oxides

The generally very high melting temperature of oxide crystals together with their very different chemical properties from those of molten semiconductors dictate significant technological differences in oxide crystal pullers as compared to semiconductor ones. Firstly, for the higher melting point oxides, precious metal crucibles (notably platinum and iridium) are required to contain the melts. For some lower melting point multi-component oxides, molybdenum or tungsten crucibles are satisfactory, however.

Iridium crucibles have been utilised for the growth of $MgAl_2O_4$ (melting point 2105 °C) and for sapphire (2050 °C). For materials melting at yet higher temperatures, RF coupling to the melt contained in a cold crucible has been employed [32].

Again because of the high temperatures involved, extensive and efficient thermal insulation of the pulling chamber is required. This is provided by refractory materials such as alumina, magnesia, zirconia and thoria. The melting point of the refractory must of course exceed that of the crystal being grown, which requirement exludes the use of alumina for some applications.

To prevent reduction of the melt it is necessary, with some materials, to provide a partial pressure of oxygen within the growth environment and then one runs into problems of oxidation of the precious metal crucibles. This can give rise to particles of, for example, iridium oxide which are wafted into the flowing ambient gas and which subsequently settle onto the melt surface and become incorporated into the crystal [33]. Careful control of oxygen partial pressure and forced gas flow configurations together with the use of baffles are necessary to minimise this effect. Where oxygen is necessary to maintain the melt stoichiometry, the outside of the crucible can be flame-sprayed with zirconia; this increases the crucible life several fold.

Radio frequency induction heating provides the cleanest and most readily applicable method for heating the precious metal crucibles although, in the interests of economy, resistance heating is sometimes employed particularly for lower melting point materials. Radio frequency heating can however bring its own problems. For example, some refractories such as zirconia tend to couple into the RF field at high temperatures producing localised hot spots which can result in crucible melting. Even pure zirconia can behave in this way if its properties have become modified by spillage of the melt components during charging of the crucible.

Where serious problems arise from distortion of the precious metal crucibles under use, this is believed to be most likely due to a relative expansion occurring between crucible and solidified material [35]. It is exacerbated by melts which effectively wet the crucible material such as, for example, garnet melts which tend to wet iridium crucibles.

The configuration and mounting of seed crystals requires special consideration. Long seeds are necessary with very high melting point oxides in order to avoid thermal degradation of the seed holder and pull rod. Where such seeds are not available, initial growth can be performed by attaching capillary tubes made of the crucible material to the pull rod which, when dipped into the melt, draw up and directionally solidify a column of melt which acts as a seed [36].

It is much less common to rotate or translate the crucible in oxide growth and accordingly the rate of growth is in excess of the pulling speed because of the rate of fall of liquid level in the melt. For stable growth one is limited to crystal diameters not significantly in excess of half of the crucible diameter.

Because many of the complex oxide systems of electronics interest, such as garnets and spinels, do not correspond to a congruently melting composition it is necessary to induce growth only at a very low rate in order to give time for diffusion away from the interface of the excess component. Failure to do this leads to dramatic degradation in the perfection of the crystals due to the occurrence of constitutional supercooling (described in Chap. 9). Typical growth speeds are of the order of a few millimetres per hour or even less. To reduce the thermal gradients in the cooling crystal and, thereby, to minimise the locked-in thermal stress (which, in severe cases and with specific materials, can result in catastrophic cracking of the crystal), after-heaters, both passive and active, are employed. In consequence growth runs are very long, commonly extending to more than one week. This dictates that the process be automated and automatic diameter control using a weighing technique was introduced extensively into the industry in the early 1970s. This is discussed in detail in Chap. 8.

The very high temperatures encountered necessarily result in high temperature gradients and in consequence, a very strong buoyancy-driven convection which is turbulent in character. This turbulence gives rise to a banded distribution of both solute concentration and varying stoichiometry in the crystal and is deleterious in many device applications. Worse still it has been found that the coupling between the natural buoyancy-driven flow and the flow due to the rotating crystal can result in catastrophic changes in flow pattern under some conditions which produce very marked temperature changes in the melt beneath the crystal resulting in dramatic changes in diameter or, in some cases, to the melting-off of the crystal with it subsequently shattering as a result of the thermal shock which it has so received [37]. This is particularly true of the material bismuth silicon oxide. Understanding and controlling convection in oxide crystal growth is a vital pre-requisite to the attainment of high perfection large diameter crystals and this aspect is considered in detail in the following chapter.

Fig. 3.9. 'Autox' puller for oxide crystal growth (courtesy of Metals Research Semiconductors Ltd.)

Virtually all the commerically important oxide single crystals are currently grown by the Czochralski technique. (The major exception is quartz which is grown hydrothermally). An automated commerical puller for the growth of high melting point oxides is shown in Fig. 3.9.

4 Convection and Flow in the Melt

4.1 Melt Convection

4.1.1 Introduction

The importance of melt flow in controlling dopant incorporation and distribution on both macroscopic and microscopic scales was made evident in the previous chapter for the cases of silicon, gallium arsenide and oxide crystal growth.

Melt growth processes tend to be thermal-transport controlled because the high temperatures involved readily generate high thermal gradients which are the driving force for both conductive and convective heat transport. Moreover, the high temperatures also tend to mean a low undercooling of the interface due to kinetic processes because of the Arrhenius dependence of the latter on temperature. Thermally-driven convection (both buoyancy-driven flows and, in processes where there is a free melt surface, thermo-capillary flows) is therefore a controlling factor in melt growth. The low Prandtl number (Pr) of molten semiconductors means that heat transport in these materials is in a mixed regime where both conductive and convective contributions are important. (Pr = v/κ where v is the kinematic viscosity and κ the thermal diffusivity). The onset of time-dependent fluid motion is readily observed in such fluids giving rise to dopant striations [38].

The flow in a Czochralski crucible is dauntingly complicated to elucidate. At the heated vertical wall of the crucible the melt is made buoyant, rises and is then turned inwards radially at the melt surface. The rotating crystal acts as a centrifugal fan sucking up fluid axially, spinning it up in a thin Ekman layer (see below) of thickness $(v/\Omega)^{1/2}$ and ejecting it tangentially (Ω is the crystal rotation rate). The radially outward flow due to the rotating crystal meets the radially inflowing fluid driven by the hot vertical crucible wall and a downflow occurs at some radial distance which depends on the relative strength of the crystal flow and the buoyancy-driven convection. Several additional factors complicate this behaviour. Firstly the radial temperature gradient across the melt surface will lead to a thermo-capillary component of flow [39]; the description of this is complicated by the fact that the melt surface is not planar since the growing crystal supports a liquid meniscus which is typically several millimeters in height. Secondly the loss of heat from the melt surface produces a destabilising vertical temperature gradient in the upper region of the melt which can give rise to a further mode of convection which is described in Sect. 4.3.1.

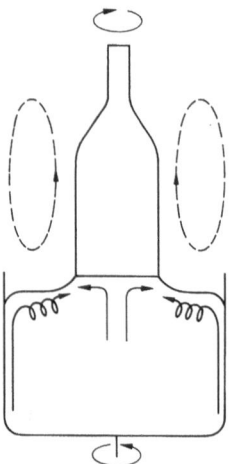

Fig. 4.1. Schematic representation of natural and forced convective flow in a crystal puller

Rotation of the crucible – commonly employed in the growth of semi-conductor crystals–can have a rather potent effect on the strength of the buoyancy-driven flow. More than that, the combined effects of crystal and crucible rotation, even in an iso-thermal fluid, are complex and depend on whether the two are iso-rotated or counter-rotated. Under some circumstances with iso- (but differential) rotation a Taylor-Proudman cell can be established beneath the crystal with a detached shear layer separating it from the outer region which rotates (approximately) as a solid body with the crucible [40], [41]. (see Fig. 4.9). Counter-rotation yields a more complex flow pattern beneath the crystal with a stagnation layer separating the flows driven by crystal and crucible rotations respectively. Finally coupling between the convective flow in the melt and in the ambient gas (or, in the case of the liquid encapsulation Czochralski technique, the encapsulant) can be significant. These several processes are illustrated schematically in Fig. 4.1.

Clearly then, before we can understand the incorporation of dopant into a Czochralski crystal, we must first have a detailed understanding of the melt flow behaviour. To do this, we have to solve the second order non-linear partial differential equation describing flow in an incompressible, Newtonian fluid – the Navier-Stokes equation. Since almost every paper reporting on simulation of flow in a crystal growth melt starts with this daunting equation as its "Equation (1)", it is incumbent on the crystal grower to understand something of its meaning. Accordingly in the following section an attempt is made to elicit the fundamental physical significance of the various terms in this equation and to outline its derivation.

4.1.2 The Navier-Stokes Equation

The force acting on a volume element of fluid is equal to the pressure exerted over its bounding surface. Denoting the pressure by p and the element of surface as dS we have the force $= -p\hat{n}dS$ where $\hat{n}$ is the unit normal to the element of surface. Converting this to a volume integral:

$$-\int p\hat{n}ds = -\int \nabla p dV \qquad 4.1$$

Equating this force to the mass acceleration product $\int\rho(dv/dt)dV$ we have

$$\rho dv/dt = -\nabla p \qquad 4.2$$

If the fluid is additionally acted on by some external force, then a force term must be added. The most commonly encountered force is gravity imposing a mass acceleration of ρg, where g is the gravitational acceleration.

Thus we have:

$$\rho dv/dt = -\nabla p + \rho g \qquad 4.3$$

The substantive derivative dv/dt denotes the rate of change of the velocity of a given "particle" in the fluid. What we wish to specify is the velocity vector at each and every point in the fluid and this is given by the partial derivative $\partial v/\partial t$. These two quantities are related by the operator:

$$d/dt = \partial/\partial t + v \cdot \text{grad} \qquad 4.4$$

Substituting this into equation 4.3 yields the Euler equation:

$$\partial v/\partial t + v \cdot \nabla \cdot v = -\frac{1}{\rho}\nabla p + g \qquad 4.5$$

This is the equation of motion of an inviscid fluid. It is a non-linear, first order equation, the non-linearity existing in the inertial term $v. \nabla v$.

The Navier-Stokes equation is obtained from the Euler equation by adding a term which represents the irreversible momentum transport in the fluid caused by viscous dissipation. This is achieved by replacing the (scalar) pressure in the Euler equation by the viscous stress tensor. When this is done we obtain an additional term on the right hand side of the Euler equation which is $v\nabla^2 v$. Thus we have the Navier-Stokes equation:

$$\partial v/\partial t + v \cdot \nabla \cdot v = -\frac{1}{\rho}\nabla p + v\nabla^2 v + g \qquad 4.6$$

The equation describes momentum transport in a viscous fluid subjected to a gravitational force. To this we have to add an equation which expresses the fact that the fluid is incompressible; i.e. the divergence of the flow at every point in the fluid is zero:

$$\nabla \cdot v = 0 \qquad 4.7$$

For the study of Czochralski growth we are interested in situations where the flow is rotational about some axis (the pulling axis). The description of such a flow is much simplified by observing it from a frame of reference which rotates with the same angular velocity and about the same axis as the body so that, in this frame of reference, the boundary conditions are time-invariant if the rotation is steady or at least have the same time variation as that imposed on the rotation.

If we rotate a confined fluid body at a steady angular rate then, after a time sufficient for the decay of any initial transient, the fluid rotates as a solid body at the rate Ω so that there are no viscous stresses within the fluid. Any additional imposed forces will cause deviations from this solid body rotation and these comprise the flow as seen in our rotating frame of reference.

Adoption of a rotating frame is accomplished by the introduction of centrifugal and Coriolis forces into the Navier-Stokes equation. The appropriate transformation is:

$$dv/dt \rightarrow (dv/dt)_{rot} + \Omega \times (\Omega \times r) + 2\Omega \times v_{rot} \qquad 4.8$$

where $\rho\, dv/dt$ is the actual force experienced by the fluid particle. The centrifugal force is $\rho\Omega \times (\Omega \times r)$ and the Coriolis force is $2\,\rho\Omega \times v_{rot}$ where v_{rot} is the velocity measured in the rotational frame. Substituting Eq. 4.8 into Eq. 4.6 gives us the Navier-Stokes equation in the rotating frame:

$$\partial v/\partial t + v \cdot \nabla \cdot v - \frac{1}{\rho}\nabla p - \Omega \times (\Omega \times r) - 2\Omega \times v + v\nabla^2 v + g \qquad 4.9$$

where we have dropped the subscript 'rot'. For a non-stratified fluid (constant density) with a conservative external force (such that it can be written as the gradient of a scalar potential) we can remove the gravitational acceleration term and the centrifugal term by writing:

$$g = -\nabla\phi \qquad 4.10$$

and by defining a reduced pressure

$$P = p - \phi - \tfrac{1}{2}\Omega^2 x'^2 \qquad 4.11$$

where x' is the distance of the point from the axis of rotation. We then have:

$$\partial v/\partial t + v \cdot \nabla \cdot v = -\frac{1}{\rho}\nabla P - 2\Omega \times v + v\nabla^2 v \qquad 4.12$$

which is the required equation.

To acquire some understanding of the behaviour of the Navier-Stokes equation, consider limiting cases where only two terms are dominant. Firstly if we consider steady flows only then $dv/dt = 0$. Secondly if the flow is inviscid then the term $v\nabla^2 v$ can be neglected. This is in general valid well away from solid boundaries. For this viscous term to be negligible compared to the Coriolis term then, in terms of some typical scale velocity U and scale length L, we must have

$$U/L^2 \ll \Omega U \qquad \text{or } E = 1/\Omega L^2 \ll 1 \qquad 4.13$$

The quantity E is called the Ekman number. It is a measure of the ratio of viscous to Coriolis forces.

For the Coriolis force to dominate over the inertial force. (v.∇v) we must have

$$U^2/L \ll \Omega U \qquad \text{or} \quad R_0 = U/\Omega L \ll 1 \tag{4.14}$$

where R_0 is called the Rossby number. It is a measure of the ratio of inertial to Coriolis force.

If both $R_0 \ll 1$ and $E \ll 1$ then the Navier-Stokes equation reduces, for steady motion, to:

$$2\rho\Omega \times v = \nabla P \tag{4.15}$$

Such motion – where there is a balance between pressure gradient and Coriolis force – is called, in the geophysical field, a geostrophic motion [42]. It has the property that both the Coriolis force and the pressure gradient are perpendicular to the flow direction. This contrasts with irrotational flow where the pressure varies along a streamline. In the present case it is constant along a streamline. This property of geostrophic flow is evident in the familiar weather maps where, away from frontal systems, the wind direction is nearly parallel to the isobars, rather than orthogonal to them.

Rewritten in terms of these dimensionless numbers, the Rossby and Ekman numbers, with time scaled by Ω^{-1} and pressure by $\rho\Omega UL$, the Navier-Stokes equation becomes:

$$\partial v/\partial t + R_0 v \cdot \nabla \cdot v + 2\hat{k}v = -\nabla P - E\nabla^2 v \tag{4.16}$$

where $\hat{k}$ is a unit vector parallel to the rotation axis.
Taking the curl of Eq. 4.15 gives:

$$\nabla \times (\Omega \times v) = 0 \tag{4.17}$$

which, using vector identities, becomes:

$$(\Omega \cdot \nabla \cdot)v - (v \cdot \nabla \cdot)\Omega + v(\nabla \cdot \Omega) - \Omega(\nabla \cdot v) \equiv 0 \tag{4.18}$$

Noting that for an incompressible fluid $\nabla.v = 0$, Eq. 4.18 becomes

$$\Omega \cdot \nabla \cdot v = 0 \tag{4.19}$$

which expresses the Taylor-Proudman theorem. Thus, if we have rotation parallel to the z axis:

$$\Omega \partial v/\partial z = 0$$

i.e $\qquad \partial u/\partial z = \partial v/\partial z = \partial w/\partial z = 0 \tag{4.20}$

where u, v, w are the components of the vector v.

The theorem implies that the particle velocity is independent of the coordinate measured along the axis of rotation, i.e. the flow is purely two-dimensional. Thus if we have a rotating crucible containing a melt and we dip into it a disc (the crystal) then, to the extent that viscous effects can be ignored, a two-dimensional flow will exist under the disc, rotating at a rate intermediate between that of the disc and of the crucible while at radii greater than that of the disc the fluid will rotate as a solid body with the angular velocity of the crucible. Separating these two regions is a thin vertical shear layer in which viscous effects are dominant. This behaviour was experimentally verified by Hide and Titman [43] some years ago.

We have just seen that viscosity cannot be neglected in the presence of these shear layers. Viscosity also becomes important close to rigid bounding surfaces and the geostrophic condition and with it the Taylor-Proudman theorem break down and the flow becomes three-dimensional in order to satisfy the equation of continuity (Eq. 4.7) in this region. The thin layer adjacent to the solid boundary of a rotating fluid over which this occurs is known as an Ekman layer.

The Ekman layer adjacent to a surface which is orthogonal to the rotation axis has the property, which is not a characteristic of a boundary layer in irrotational flow, that the velocity field is uniform in the plane of the surface. This is illustrated in the following section which describes flow adjacent to a rotating disc.

4.2 Flow Due to a Rotating Disc

A rotating disc acts as a centrifugal fan, drawing in fluid axially imparting angular momentum to it within the (rather thin) Ekman layer and expelling it tangentially. Hence the flow is fully three-dimensional and account must be taken of all three components of the Navier-Stokes equation. Von Karman [44] has shown that there exist similarity solutions for the components of the velocity and pressure fields. If we non-dimensionalise the z coordinate (the rotational axis) by the Ekman layer thickness so that:

$$\zeta = z(\Omega/\nu)^{1/2} \qquad\qquad 4.21$$

then Von Karman showed that the similarity solutions were of the form

$$u = r\Omega F(\zeta)$$

$$v = r\Omega G(\zeta) \qquad\qquad 4.22$$

$$w = (\nu\Omega)^{1/2} H(\zeta)$$

$$p = \nu\Omega P(\zeta)$$

where r is the radial co-ordinate.

Substituting these into the component Navier-Stokes equations yields the following set of simultaneous ordinary differential equations in the functions F, G, H and P.

$$F^2 - G^2 + HF' = F'' \qquad\qquad 4.23$$

$$2FG + HG' = G'' \qquad\qquad 4.24$$

$$2F + H' = 0 \qquad\qquad 4.25$$

$$-HH' - 2F' = P' \qquad\qquad 4.26$$

where the prime indicates differentiating with respect to ζ. The boundary conditions on the disc surface ($\zeta = 0$) are:

$$F = H = P = 0; \qquad G = 1 \qquad\qquad 4.27$$

and in the far field ($\zeta = \infty$)

$$F = G = 0 \qquad\qquad 4.28$$

Solution of this set of equations by numerical integration was first performed by Cochran [45] whose results are shown in Fig. 4.2. The axial velocity (H) rises to a steady value well away from the disc. Close to the disc, its value can be expanded as a power series in ζ as follows:

$$F = a\zeta - 1/2\zeta^2 - 1/3b\zeta^3 + \ldots \qquad\qquad 4.29$$

$$G = 1 + b\zeta + 1/3a\zeta^3 + \ldots \qquad\qquad 4.30$$

$$H = - a\zeta^2 + 1/3\zeta^3 + \ldots \qquad\qquad 4.31$$

The tangential velocity (G) is seen to fall to about half its value at the disc's surface in the Ekman layer thickness ($\zeta = 1$). The streamlines spiral outwards so that their trace in any plane $z = z_0$ is a multi-start spiral (Fig. 4.3).

The flow due to the rotating Czochralski crystal is commonly supposed to be represented by this von Karman flow. However the latter is strictly valid only for a disc immersed in an infinite fluid medium. Recirculation of the fluid resulting from the presence of the crucible wall at a finite distance from the disc renders this approximation somewhat dubious [46] but, nonetheless the theory has proved very successful as we shall later see.

The Czochralski problem differs in a further respect from the original von Karman one in that there is a flow of fluid towards the disc resulting from the crystallisation process. If the normal crystal growth speed is v_p then the boundary condition on the disc for the normal component of flow velocity is not $w = 0$ but rather $w = - v_p$. This is equivalent to a rotating disc which is porous having a suction velocity at the disc equal to this normal crystallisation velocity. This is not to be confused with the already discussed drawing in of fluid axially

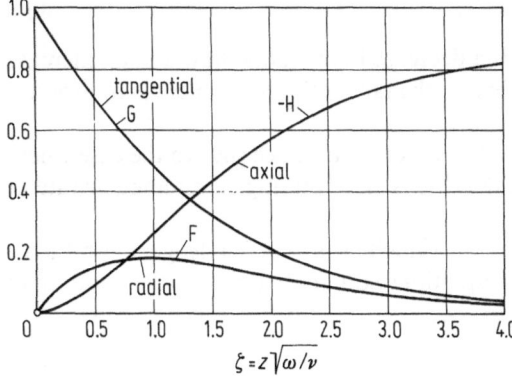

Fig. 4.2. Components of the the flow profile at a rotating disc. z is the co-ordinate normal to the disc axis and is scaled by the square root of the quotient of angular velocity and kinematic viscosity. (after Schlictling [236])

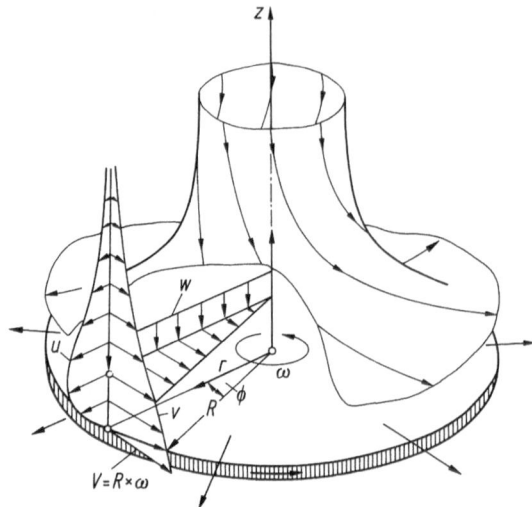

Fig. 4.3. Spiral flow at a rotating disc. (after Schlictling [236])

from infinity by the disc which is referred to as Ekman suction. Stuart [47] and others have shown that the von Karman similarity solutions can be extended to this problem where now the boundary condition on H at the disc ($\zeta = 0$) is

$$H = -a \qquad\qquad 4.32$$

where $a = w \, (v\Omega)^{-1/2}$ is the normalised suction velocity. The form of the solution is the same but the numerical value of the coefficients in the Cochran expansion about $\zeta = 0$ are changed and an additional term appears in the expansion for H.

$$F = a_1\zeta - 1/2\zeta^2 - 1/3b_1\zeta^3 + \ldots \qquad\qquad 4.33$$

$$G = 1 + b_1\zeta + 1/3a_1\zeta^3 + \ldots \qquad\qquad 4.34$$

$$H = -a - a_1\zeta^2 + 1/3\zeta^3 + \ldots \qquad\qquad 4.35$$

When suction is applied there is a decrease in the radial flow and an increase in the axial flow at infinity. The effect on the tangential component is to produce a reduction in the boundary layer thickness.

This analysis has been extended by Rogers and Lance [49] to the case where the body of the fluid remote from the disc is also rotating, at an angular rate of Ω_f.

Defining s as the ratio of these two velocities ($s = \Omega_f/\Omega$), Rogers and Lance have shown that the Von Karman equations still apply but with the addition of a term in s^2 to equation 4.23 viz:

$$F'' = F^2 - G^2 + HF' + s^2 \qquad\qquad 4.36$$

with the modified boundary condition on G in the far field:

$$G(\infty) = s \qquad\qquad 4.37$$

Stable solutions were obtained for only a limited range of values of s.

One might guess that the outward flow over the major part of the surface of a disc of *finite* radius would be uninfluenced by the "edge effect" at the disc periphery provided that the disc radius is much greater than the Ekman layer thickness. As we have seen, the Ekman layer is generally very thin so that this condition is readily satisfied and indeed the expectation that infinite disc theory should be valid appears to be well borne out in practice. In the situation that the fluid is rotating more rapidly than the disc however, so that the flow adjacent to the disc is radially inward toward the axis, the effect of the discontinuity at the disc periphery is more pronounced [50].

The limitations of these infinite disc solutions when applied to Czochralski growth should be firmly borne in mind. In addition to the problems of the inappropriateness of the boundary conditions to Czochralski growth is the question of the stability of these flows. Thin shear layers such as occur adjacent to a rotating disc are known to be subject to instability at sufficiently high rotation rates. However it would appear that these are outside the range of rotations commonly employed in Czochralski growth.

We must also consider the region remote from the disc. Hide and Titman [43] found that above a critical value of the Rossby number the flow became non axi-symmetric and they derived an empirical relationship between the critical value of the Rossby number and the experimentally chosen Ekman number. Experiments by Jones [51] on model Czochralski melts have shown that the flow in the bulk of the melt becomes non axi-symmetric at Rossby numbers of order 1 if the Ekman number is sufficiently small (see Fig. 4.4) but the mechanism of instability remains unclear.

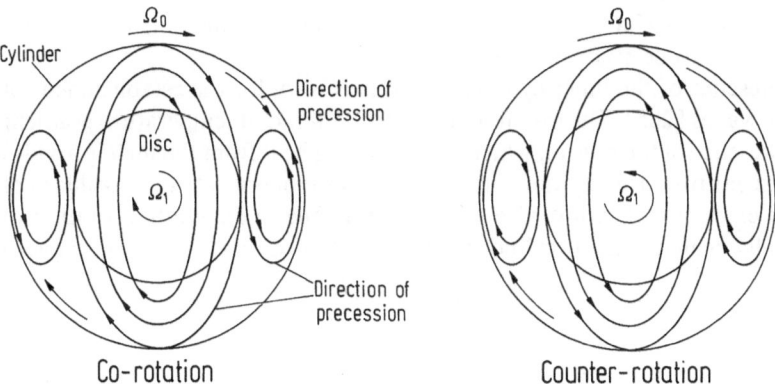

Fig. 4.4. Non-axisymmetric flow in a model Czochralski melt for co-and counter-rotation of crystal and crucible. Crystal rotation rate (Ω_1), crucible rate (Ω_0). (Jones [51])

Very recently, Kakimoto et al. [264] have experimentally observed non-axisymmetric flow in a rotating silicon melt using double X-ray radiography. The flow comprised stable eddies which rotated with the crucible which they ascribed to a baroclinic instability, well known in the meteorological field.

This arises as follows: Consider a pair of vertically mounted coaxial cylinders containing the melt in the annular space with the outer cylinder maintained at a higher temperature than the inner. The radial temperature gradient will produce an overturning motion in the fluid with hot fluid rising at the outer cylinder wall. The temperature field is stably statified and the flow is azimuthal.

If now the annulus is rotated about its axis this will produce a gyroscopic torque which tends to inhibit the meridional overturning. If the Ekman number is sufficiently small ($\ll 1$) and the Rossby number is also small, the flow in the body of the fluid will be geostrophic with the gyroscopic torque (proportional to $\nabla \times (2\,\rho\Omega \times u)$) balanced by the 'buoyancy torque' $q \times \nabla\rho$.

This geostrophic balance produces a flow vector which is nearly horizontal; i.e. the rotation largely suppresses the overturning motion. An axisymmetric mode will thus have a nearly azimuthal flow and this is very inefficient at transferring heat radially from the outer to the inner cylinder. In practice, for sufficiently small Rossby number, the flow suddenly loses axisymmetry as the rotation rate is increased through a critical value. In the non-axisymmetric regime the flow resembles the atmospheric 'jet stream' with eddies carried around the annulus in the direction of its rotation. The weaving of the stream between the inner and outer cylinders provides a much enhanced mechanism for heat transfer between the two cylinders. At yet higher rotation rates the eddies are seen to become unsteady (and are said to 'vaccillate'). Such behaviour is sometimes referred to as 'geostrophic turbulence' [268].

The above is a very simplified description of complex phenomena which comprise a number of regimes collectively described as 'sloping' convection and which have been comprehensively reviewed by Hide and Mason [265]. The mathematical description of even the axi-symmetrical state is formidable and only careful interplay between theory and experiment has led to the elucidation of the operative mechanisms (Eady [266]). Its exact relevance to Czochralski growth has yet to be studied in detail.

The presence of the inner cylinder is not necessary for the establishment of baroclinic waves [265]. The essential ingredient is a radial temperature gradient and the presence of the crystal, which is abstracting heat from the melt, provides this. The shear produced by the differential rotation of crystal and crucible is of course an extra complication. The effect of shear due to differential rotation of differentially heated annular cylinders has been studied by Lambert and Snyder [267].

4.3 Natural Convective Flow

4.3.1 Buoyancy-Driven Flow

Czochralski crucibles are usually heated from the side and the hot layer of fluid adjacent to the vertical crucible wall is therefore buoyant and rises entraining fluid which, in the absence of crystal and crucible rotation, is carried around in a toroidal motion giving rise to a single "doughnut" circulation with fluid descending beneath the axis of the crystal [52]. The driving force for this motion is expressed by the non-dimensional Grashof number:

$$Gr = \alpha g \Delta T d^3 / \nu^2 \qquad 4.38$$

where a temperature difference ΔT exists across some characteristic distance in the system d. To describe this motion, we must further extend the Navier-Stokes equation to take account of the driving force provided by this buoyancy. To do this we have to take account of the spatial variation of the density. The local density is related to the local temperature through the linear phenomenological law:

$$\rho(T) = \rho(T_0)\{1 + \alpha(T - T_0)\} \qquad 4.39$$

where α is the thermal expansion coefficient. This introduces temperature as a variable into the Navier-Stokes equation so that it is now coupled to the heat transport which, as we shall see, is expressed by a second-order differential equation which contains both the temperature and the velocity. To simplify this coupling it is common to make the Boussinesq approximation [53] which assumes that the various coefficients such as thermal conductivity, density and kinematic viscosity are constant (i.e. temperature invariant). The one exception made is the density in the term ρg because the acceleration resulting from a change in ρ which is:

$$\delta \rho g = \alpha \delta T g \qquad 4.40$$

can be quite large compared to other terms in the equation. The form of the Navier-Stokes equation in a rotating frame is then:

$$\partial v / \partial t + v \cdot \nabla \cdot v = -\frac{1}{\rho_0} \nabla P + \nu \nabla^2 v + (1 + \alpha \delta T) g - 2 \Omega \times v \qquad 4.41$$

where $\qquad \rho_0 \equiv \rho(T_0)$ and $\qquad \delta T = T - T_0$

The heat transport (energy conservation) equation is:

$$\partial T / \partial t + v \cdot \nabla T = \kappa \nabla^2 T \qquad 4.42$$

where κ is the thermal diffusivity.

Solution of these coupled transport equations in the domain of a Czochralski melt requires considerable computing power. A number of simulations have been performed by various workers who typically have idealised the configuration in the manner shown in Fig. 4.5. The conditions on the temperature at the crucible wall can either be expressed as isothermality or as a fixed heat flux through the walls with, perhaps, an insulating base. The most serious idealisation is the location of the crystal-melt interface in the plane of the bulk melt surface i.e. ignoring of the presence of the meniscus supported by the growing crystal. The value of such simulations is that they give an indication of the flow patterns which can occur in the melt under various combinations of crystal and crucible rotation and imposed thermal gradients [54] and can locate regions of parameter space where bifurcations to new flows can occur [41]. Most of the simulations to date have assumed axi-symmetry but three-dimensional simulations are now becoming more common [52], [55], [56]. Given the experimental studies of Jones cited above [51], the importance of three-dimensional modelling is clear.

The other major limitation of such simulations is that they prescribe the shape of the crystal-melt interface whereas, in practice, this is determined by the heat transport in the combined system:- melt, crystal and ambient gas. This problem has been addressed in recent simulations and is considered below in Sect. 5.3.

For the moment we continue to focus on purely thermo-hydrodynamic aspects of the melt and review two characteristic features of oxide crystal growth, namely (a) the appearance of a radial spoke pattern on the surface of most oxide melts and (b) a catastrophic flow transition observed in some melts: gadolinium gallium garnet and bismuth silicon oxide in particular.

The visual appearance of an oxide melt from which a crystal is growing is characterised by a radial pattern of dark spokes which precess with the crystal

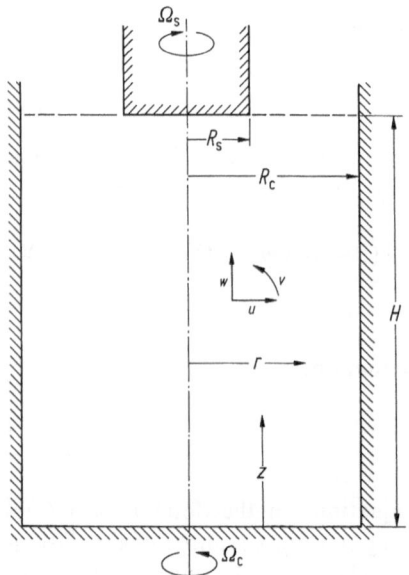

Fig. 4.5. Simulation of flow in the melt; typical geometry employed. (Langlois [54])

rotation [57]. Probe thermocouples reveal that the dark spokes are relatively cold and hence delineate sheets of descending fluid. Miller and Pernell [58], and Jones [59] have independently demonstrated this phenomena in a radially heated cylindrical crucible containing water as a model fluid. It has been established that the structure is confined to the near-surface region of the melt and that beneath this there is a large thermally-mixed region with a stagnant layer having a small, but stabilising, vertical temperature gradient. Between each spoke, fluid spirals inwards (Fig. 4.6).

Jones [60] has established a plausible model for this behaviour. He measured a destabilising vertical temperature gradient from the near- surface layer and showed experimentally that the spoke pattern appeared when the Rayleigh number:

$$Ra = Gr\ Pr \hspace{6cm} 4.43$$

exceeded a critical value which lay in the range $230 < Ra < 290$. This thin layer of unstably-stratified fluid results from the rapid loss of heat from the melt surface by the buoyant fluid which has risen up the crucible wall and is flowing inward towards the crystal. Jones [60] carried out a linear perturbation analysis from which he was able to predict a critical Rayleigh number for the onset of a convective overturning which had the form of the spoke pattern. His predicted critical value is $Ra_c = 255$ in very good agreement with the measured value.

Certain oxide melts, which have in common that they are rather viscous, can induce a catastrophic instability in the growth of the crystal as the crystal diameter and/or the crystal rotation rate are increased [61]. Dark "lobes" appear on the melt surface attached to the crystal which grow and spread radially (Fig. 4.7) obliterating the radial spoke pattern until it extends over the whole melt surface which then becomes featureless. Readings from thermocouples placed beneath the growing crystal record a sudden rise in temperature

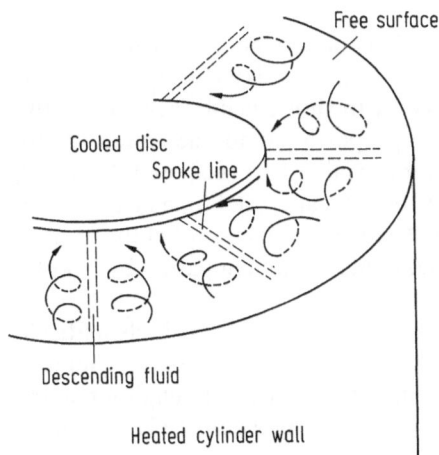

Fig. 4.6. Schematic representation of spoke pattern in an oxide melt. (Jones [59])

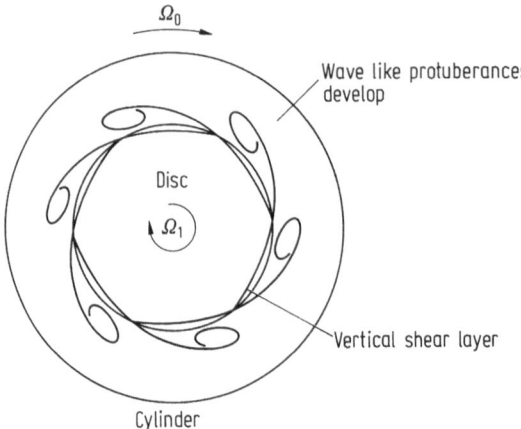

Fig. 4.7. Schematic representation of 'lobe' instability in a oxide melt. (Jones [46])

of several tens of degrees [62], [63] which causes the meniscus height to increase markedly producing, in the case of bismuth silicon oxide, a static instability where the meniscus breaks at the neck. The crystal is now suddenly disconnected from its heat source (the melt) and the large thermal shock which it suffers as a result causes it to shatter, the pieces falling spectacularly into the melt. With gadolinium gallium garnet the effects are commonly less dramatic [64]; the meniscus does not break but significant melting-back of the previously very convex crystal-melt interface occurs. Subsequent growth takes place with a near-planar interface. This effect has been used to advantage to control the defect structures in the material. [64]. (Fig. 4.8). Specifically this transition, when it occurs, results in the disappearance of a core facet in the crystal which gave rise to a high elastic strain. The growth following the transition occurs on a slightly concave interface so that the facet is absent and the strain very much reduced.

The nature of this melt instability is not fully understood. It occurs when the effect of the "rotating disc" flow due to the crystal exceeds some critical value in relation to buoyancy-driven flow generated by the hot crucible walls. These two components of flow will be respectively proportional to the square of the Reynolds number and to the Grashof number. Carruthers [65] used this simple semi-quantitive argument to predict that the critical crystal diameter at which the transition would occur would vary as the inverse square root of the rotation rate. These arguments have been examined in more detail in papers by Nikolov et al. [66].

The various boundary layers present in the melt are indicated schematically in Fig. 4.9. The daunting complexity alluded to at the beginning of this chapter is now evident! Resolution of the flow structure by numerical simulation requires a fine mesh throughout the melt volume which, in turn, requires a large amount of computing power.

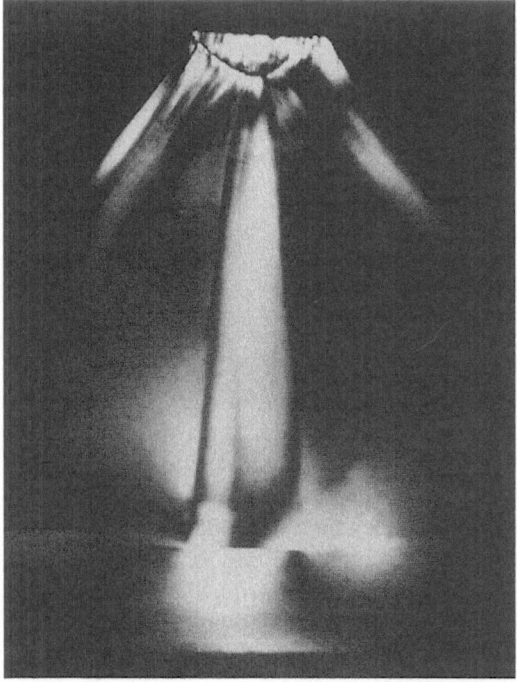

Fig. 4.8. Optical micrograph of a longtitudinal section through a gadolinium gallium garnet single crystal. Striations reveal the interface shape. Note the disappearance of the central facet when interface melt-back occurs. [251] (Mag. × 4)

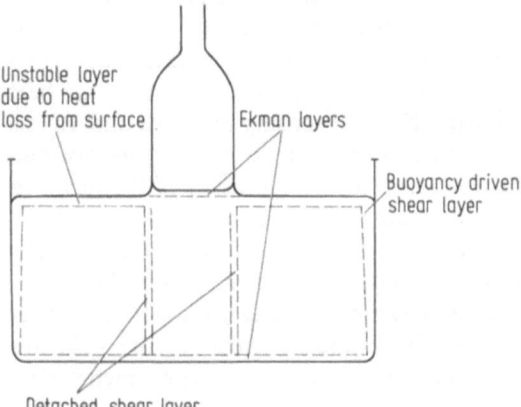

Fig. 4.9. Schematic representation of the momentum boundary layers present in a Czochralski melt

4.3.2 Marangoni Flow

The surface tension of most liquids is temperature dependent and, since there exists a radial temperature gradient in a melt from which a crystal is growing, there will exist a radial gradient of surface tension in the melt surface. This will produce a surface traction which will induce a circulation within a thin layer of melt adjacent to its free surface. Miller and Pernell [58] suggest this as a mechanism producing the spoke patterns but this has been discounted by Jones [59]. A number of simulation experiments have been performed by Schwabe and Metzger [67] which appear to demonstrate that Marangoni flow in a Czochralski melt is very significant. This conclusion is supported by a number of numerical simulations [68].

4.3.3 Convective Turbulence

As the driving force (control parameter in the jargon of dynamical systems theory) for convective flow (the Grashof or Marangoni number) is increased, a point is reached at which the flow becomes time-periodic. This transition occurs most readily (i.e. at the lowest value of the control parameter) for low Prandtl number fluids such as liquid metals. Further increase in the control parameter results, in general, in progressively more complex time-dependence until the flow is more properly described as turbulent. This evolution may not be continuous and indeed Müller and Neumann [69], using a centrifuge to increase the buoyancy force and hence the Grashof number, have shown that a transition back from time-periodic to stationary flow can occur with some molten metals. However, further increase in the parameter leads to a progressively more complex flow of steadily increasing amplitude. The importance of the Coriolis force in stabilising stationary flows has recently been recognised [256].

The classical concept of the transition to turbulence, due to Landau and Lifshitz [70], that full turbulence is reached through an infinite sequence of instabilities each contributing a new (incommensurate) frequency, has been replaced by the concept of deterministic chaos. According to Ruelle and Takens [71], a chaotic state emerges as a result of a bifurcation of an oscillatory state into a strange attractor. This concept arises in consideration of the temporal dynamics of non-linear systems (see for example Thompson and Stewart [72]). The spatio-temporal dynamics are incomparably more complex but, in systems where the number of spatial modes is severely restricted (such as in the famous Lorenz model [73]), a few distinct and separate routes to chaos have been identified. These depend not only on the Prandtl number of the fluid, but sensitively on the boundary conditions and in particular on the geometry of the fluid container [74].

In configurations where the aspect ratio of the fluid differs markedly from unity and the flow consists of a large number of convective cells, the behaviour is less obviously regular and well-defined possibly due to the existence of "structural" defects in the cellular pattern which induce and permit the propagation of vertical vorticity. Czochralski melt geometries are, with a full crucible, of unit order aspect ratio. However, as growth proceeds, that aspect ratio gets progressively smaller and the convective flow will in general pass through a large number of bifurcations in the duration of the growth cycle so that frequent changes in the spectral content of the melt temperature fluctuations are to be expected. The effect which this has on microsegregation in the crystal is considered in Chap. 6.

Temperature oscillations in Czochralski melts have been studied by a number of workers. Early work is reviewed by Muhlbauer et al. [252] ; the first author further calculated the effect of the fluctuations on segregation coefficient values (Muhlbauer [253]). A more comprehensive treatment of the theory has subsequently been given by Wheeler [254]. Theory and experiment has been comprehensively reviewed by Müller [262].

The effect of temperature fluctuations on oxygen microsegregation and swirl defect formation in Czochralski silicon was studied by Murgai et al. [255]. An experimental study of calcium tungstate and calcium fluoride melts was made by Cockayne and Gates [239]. Striation spacings were shown to be compatible with the observed fluctuating temporal spacing. Crystal rotation produced a near-periodic oscillation of increased amplitude which correlated strongly with striation spacing.

5 Heat Transport

5.1 Introduction

In a later chapter we progress to asking questions about the stability of the growth process and to this end we need to understand the heat transfer processes throughout the crystal pulling assembly. We have seen that numerical simulation provides physical insight into the nature of the flows in the melt and their coupling to the thermal boundary conditions but that to deal with questions of stability one must properly take account of the crystal-melt interface as a thermodynamic phase boundary whose shape represents the solution of the problem and not a boundary condition. Such problems are known as Stefan problems (see for example Ockendon and Tayler [79]). First, though, we determine the temperature field in the cooling crystal.

5.2 Heat Loss from the Cooling Crystal

Latent heat generated at the crystallising interface is conducted into the crystal to be removed by radiation and convection from its surface. Where the thermal conductivity of the crystal is large, significant heat loss through the seed crystal can also occur, particularly in the early stages of growth. The measured temperature distribution in a germanium crystal, its melt and its ambient is shown in Fig. 5.1.

Mathematically, the cooling Czochralski crystal can be represented as a cylinder losing heat by (a) axial conduction (b) surface heat loss by radiation and (c) convective transfer to an ambient gas. The lower surface of this rod is the isotherm $T = T_M$ the melting temperature. In the general case this problem can be solved only by numerical integration but we can obtain some physical insight by seeking approximate pseudo-one-dimensional solutions valid for relatively long crystals and small values of the Biot number. Such a solution has been derived by Brice [80]. (The Biot number $Bi = Hr/k_s$ is a measure of the ratio of the radial to axial heat transfer. H is the surface heat transfer coefficient and r the crystal radius.)

Now the Peclet number $Pe = v_p r/\kappa_s$ for a growing crystal is very small. (Pe is a measure of the ratio of the advected to conducted heat). Accordingly, the temperature distribution in a cooling crystal will be given to good accuracy by solution of the steady state heat conduction equation (the Laplace equation),

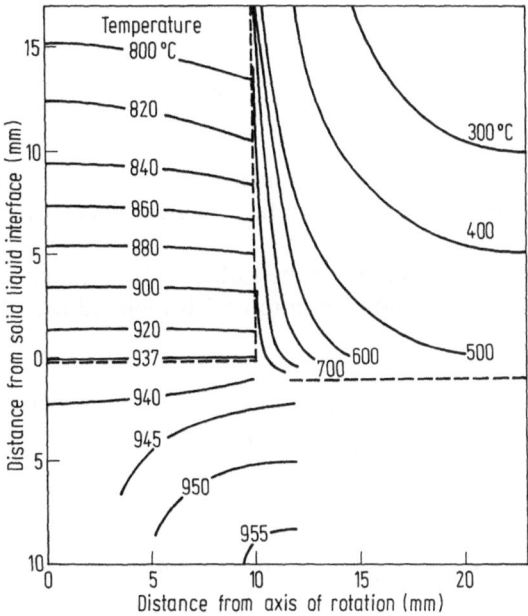

Fig. 5.1. Temperature distribution in a germanium pulling system compiled from measurements during the growth of several crystals and in the gas phase. (Brice [80])

written in cylindrical polar co-ordinates:

$$d^2T_s/dz^2 + 1/x \, dT_s/dx + d^2 T_s/dx^2 = 0 \qquad\qquad 5.1$$

where x is the radial and z the axial co-ordinate. The relevant boundary conditions are that the crystal-melt interface (the plane $z = 0$) is at the melting temperature T_M and that the cylindrical surface of the crystal loses heat at a rate characterised by the heat transfer coefficient H viz:

$$dT_s/dx + H(T - T_{amb}) = 0 \qquad 0 < z < L; x = r \qquad\qquad 5.2$$

where T_{amb} is the ambient temperature and L the crystal length. Integrating Eq. 5.1 subject to this boundary condition and in the limit of an infinitely long crystal, Brice obtains:

$$T_s - T_0 = (T_M - T_{amb}) \, 2H/r \sum_{n=0}^{\infty} \frac{J_0(\alpha_n x)}{(H^2 + \alpha_n^2)J_0(\alpha_n r)} \exp(-\alpha_n z) \qquad\qquad 5.3$$

where the α_n are the positive roots of

$$\alpha_n J_1(\alpha_n r) + HJ_0(\alpha_n r) = 0 \qquad\qquad 5.4$$

and where J_0, J_1 are Bessel functions.

Typically $H < 1$ cm^{-1} so that Eq. 5·3 can be approximated to:

$$T_s - T_{amb} = (T_M - T_{amb}) \frac{(1 - Hx^2/2r)}{(1 - Hr/2)} \exp\{-[2H/r]^{1/2}z\} \qquad 5.5$$

The above analysis applies for a planar crystal-melt interface only. The effect of interface curvature on the temperature distribution has been considered by Borodin et al. [85].

For high melting point materials, heat loss from the surface will be dominated by radiative transfer so that the boundary condition 5.2 is replaced by:

$$k_s \, dT_s/dx + \varepsilon\sigma(T_s^4 - T_{amb}^4) = 0 \qquad 5.6$$

where ε is the emissivity of the surface and σ is Stefan's constant. However, since $T_s \gg T_{amb}$, Eq. 5.6 can be approximated to the form of Eq. 5.2 by linearising to give:

$$H = 4\varepsilon\sigma T_{amb}^3/k_s \qquad 5.7$$

where now the heat transfer coefficient is a function of temperature and hence of the axial co-ordinate z.

We have assumed that the crystal is radiating into free space at some low temperature such that re-radiation back to the crystal is negligible. The reality of a crystal growing chamber is more complicated and, for high melting point materials at least, a proper account of the temperature distribution in the cooling crystal must include the radiation environment in which the crystal finds itself. This is a very complex problem which has been addressed by Stern [83], by Brown and co-workers [84] and by others. Particularly complicated is the treatment of the conical section of the crystal. This has been attempted by Stern who has used view-factor mapping together with approximate analytical representation of the maps in order to construct a complete picture of the radiation equilibrium within the puller chamber. A detailed review of heat transfer in Czochralski growth has been given by Kobayashi [82]. Measured temperature profiles in a germanium crystal pulling system are shown in Fig. 5.1.

5.3 The Thermal-Capillary Model

In Chap. 2, it was shown that maintainance of growth at a constant radius required that the three-phase boundary (which lies approximately on the isotherm corresponding to the freezing temperature) be located at a height h_0 above the bulk melt level where h_0 was determined from solution of the Laplace-Young equation for a crystal of radius r and a contacting angle Θ_L^0 ($= 13 \pm 1°$ for Ge and $11 \pm 1°$ for Si). If the heat transfer processes locate the isotherm through the three-phase boundary at some height different from h_0 then $\Theta_L \neq \Theta_L^0$ and the crystal radius will change as growth proceeds.

Modelling of the evolution of the crystal radius proceeds as follows: Suppose that the temperature field, meniscus and crystal radius (r_0) are known at some initial time t_0. After growth has proceeded for a time δt the radius will have evolved to a new value $r_0 + v_p \tan(\Theta_L - \Theta_L^0)$. The equations for the temperature fields are now solved (if melt convection is ignored then the enthalpy method [81] can be used) assuming that the profile of the meniscus is that which it had at time t_0. By locating the freezing point isotherm, the new meniscus height h ($t_0 + \delta t$) is obtained. An approximate analytical solution of the Laplace-Young equation (see Hurle [18]) can be used to calculate the new contacting angle Θ_L ($t_0 + \delta t$) and hence the new meniscus profile. This is now substituted into the boundary conditions on the temperature field and a new estimate of the position and shape of the crystal-melt interface obtained. Iteration of this procedure is continued until the successive solutions converge. Time is then incremented to $t_0 + 2\delta t$ and the procedure repeated.

5.4 Fluid Motion in a Liquid Encapsulant

The additional factors to be considered in the LEC growth of compound semiconductors are the flow and heat transfer processes which occur within the encapsulant. Hicks [86] has shown that the flow in the encapsulant is extremely weak corresponding to a Grashof number typically less than 10^{-1}. This gives rise to meridional flow velocities of the order of only 10^{-3} cm/s. The flow is therefore predominantly azimuthal, driven by the differential rotation of the crystal and crucible. Thus the radial heat flow from the crystal is predominantly by conduction in accord with the assumptions made by Derby and Brown [87] in their quasi-steady state thermal-capillary model described below. It is contrary to the suggestion made by Jordan [88] that heat transfer between crystal and encapsulant is dominated by strong convective motion in the encapsulant.

5.5 Global Heat Transport

Modelling of the time evolution of Czochralski growth using a thermal-capillary model as outlined above was carried out Crowley [89] and extended subsequently in collaboration with Stern [90]. This provided valuable information on the dynamic response of a crystal-melt interface to temperature changes in the bulk of the melt but the modelling was confined to the immediate vicinity of the meniscus and took no account of bulk convective processes.

Following this pioneering work, Brown and co-workers [68] have made an extensive modelling study of this, so-called TCM (thermal-capillary model) to include heat transfer through the crucible, convection in the melt, heat transfer through the meniscus and from the cooling crystal to a convecting gas ambient. However, the computing demands made by such a model are such that it has not yet been possible to model the convective flow up to realistic values of the Grashof number. Similar global modelling has been performed by Dupret and co-workers [91].

Atherton et al. [84] have modelled the effects of diffuse-grey radiative heat transport on silicon Czochralski growth using a view factor matrix evaluation algorithm. Neglecting convective heat transport in the melt and assuming that the crucible base was thermally insulating, they demonstrated that lowering the axial temperature gradient in the crystal resulted in a very concave-to-the-melt interface, essentially because the liberated latent heat had then to be removed radially by radiation from the crystal surface rather than by axial conduction.

For more typical axial gradients and for low growth rates (such that the rate of generation of latent heat was not a significant factor), the interface was convex-to-the-melt, the degree of convexity decreasing with increasing growth speed.

Allowing some heat loss from the base of the crucible served to decrease the convexity of the interface but could lead to a catastrophic instability – viz a supercooled melt with the crystal touching the crucible base –in the later stages of growth when the melt volume had become small. (Fig. 5.2).

Crystal pulling is a batch process which therefore does not have a true steady state – i.e. the melt volume, crystal length etc. are continuously changing with time. The above reported simulations are referred to by Brown and co-workers as the quasi-steady-state model which supposes that the characteristic thermal response times of the system are short compared to the time for significant changes in the boundary conditions (due to the batch nature of the process) to have occurred. The whole growth process can then be simulated by a sequence of quasi-steady-state snapshots.

At this point careful consideration must be given to the purpose of the modelling. If the pulling speed and heater temperature remain fixed, then the crystal radius will in general change progressively throughout the simulated growth run. However, if the purpose is to simulate growth at some specified diameter, then, by iterating the solutions at each interval, varying the heater temperature until the required radius is obtained, the heater temperature schedule required to achieve a crystal of the demanded diameter can be predicted. However this assumes that the obtained solutions are *stable* and this is considered in Chap. 8. Suffice it to say for the moment that Atherton et al. [84] demonstrated the stability of their solutions.

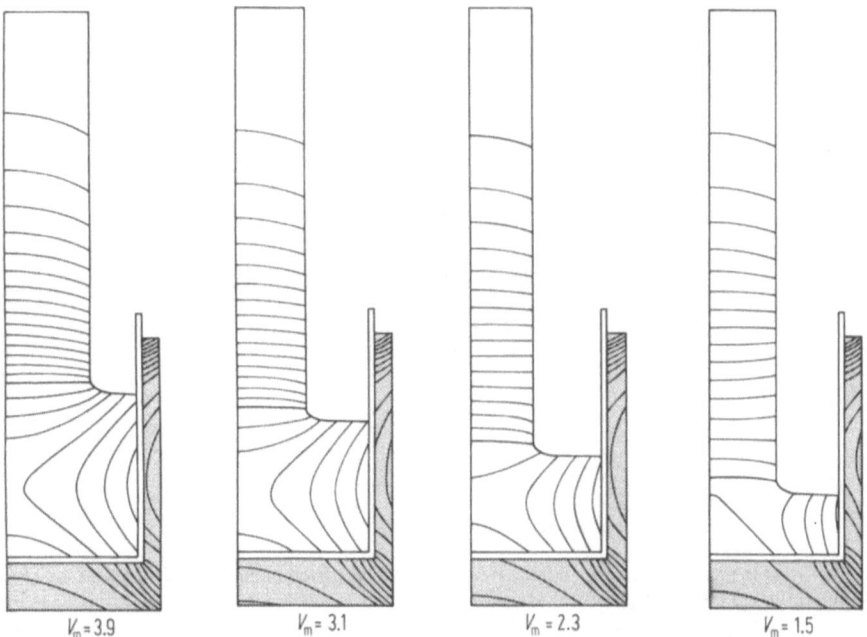

Fig. 5.2. Numerical simulation of the temperature distribution in a silicon crystal puller at progressive stages in the growth cycle. Cooling of the crucible base is shown in this simulation which results in the interface becoming concave in the later stages of growth with supercooling of the melt at the crucible base beneath the crystal. (Atherton et al. [84])

In practice, of course, the crystal radius is servo-controlled and the ultimate simulation would include within it the dynamics of the servo-controller. A simple approach to this has been made by Atherton et al. [84].

Modelling the LEC process brings with it the additional complication of properly accounting for the heat transport in the encapsulant. As explained in Sect 5.4 above, convection in boric oxide can be neglected. More difficult to model is its radiative heat transport properties. Measurements by Ostrogorsky et al. [92] have shown that clear dry boric oxide is transparent only below a wavelength of about 2 microns. (Fig. 5.3). Progressive contamination of the encapsulant during growth can be expected to make it even less transparent. The degree of thermal transparency assigned to the encapsulant can have a profound effect on the model predictions (see Bötkler et al. [93]).

Crowley et al. [90], have modelled growth of the cone section of an LEC crystal, assuming the boric oxide encapsulant to be opaque with an emissivity of 0.75. They showed that to achieve planar isotherms in the cooling crystal (a desirable condition for reducing thermal stress − see Chap. 10) one requires an encapsulant with a high heat transfer coefficient for the conical section of the

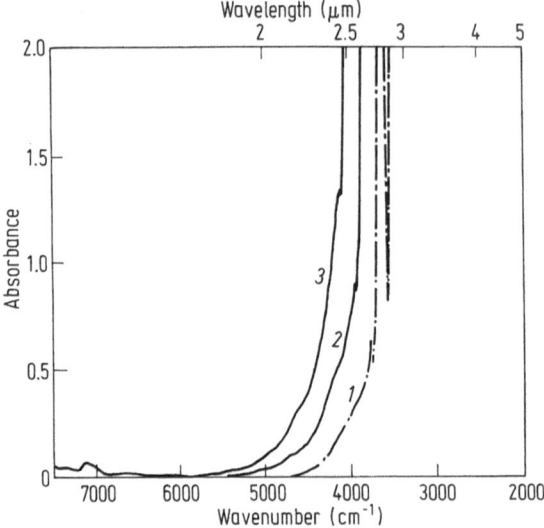

Fig. 5.3. Absorbance spectra of dry boric oxide at 1250 °C measured on films of thickness 0.2 cm (*1*); 0.4 cm (*2*); 1.0 cm (*3*). (Ostrogorsky et al. [92])

growth but with a low value of that parameter for the cylindrical portion!. Encapsulant depth is also important because a significant discontinuity in thermal gradient occurs at the crystal-encapsulant-ambient gas triple boundary.

The importance of interface shape has already been alluded to; this is sensitively dependent on melt flow conditions as has been indicated. Global modelling, to be really useful, must therefore be pursued to the point that the flows are modelled at realistic Grashof number. If the objective is to study how the crystal cools, the alternative is to take experimentally determined interface shapes as was done by Crowley et al. [90].

5.6 Optically Semi-Transparent Crystals

Some crystals are semi-transparent to thermal radiation. This is particularly the case for sapphire and a number of other oxide crystals. Cockayne et al. [237] grew a range of DyAG-YAG solid solution crystals under near-identical conditions and showed that the interface shape became progressively more convex as one went from YAG to DyAG. YAG is transparent in the 'thermal' 1–6 μm range whereas DyAG absorbs radiation extensively in this wavelength range due to the presence of the Dy^{3+} ion. Because of its near-planar interface DyAG alone was free from a strained and facetted core region. Profiles of the obtained interface shapes are shown in Fig. 5.4.

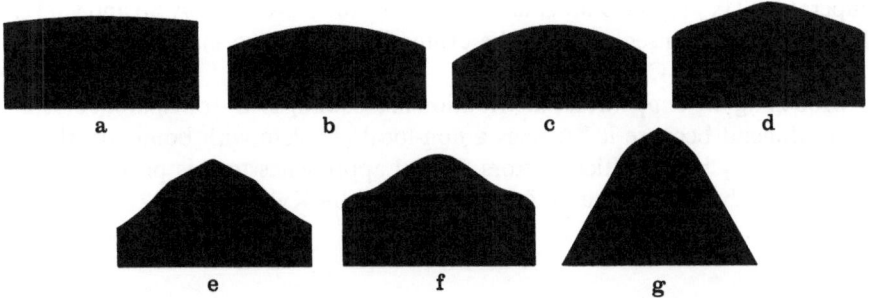

Fig. 5.4. Interface profiles for DyAG-YAG single crystals pulled under identical conditions, **a** to **g** corresponds to composition changing from 0% to 100% YAG. (Cockayne et al. [237])

Table 5.1. Transition temperatures to transparency, $T(x)$

Material	n^a	Melt Pt. K	Band Gap eV	$T(x)$ K	$T(x)/Tm.$
Si	3.4	1693	1.11	850	0.5
Ge	4.0	1210	0.67	450	0.37
GaAs	3.2	1553	1.43	1125	0.72
GaSb	3.7	978	0.69	515	0.53
InP	3.0	1343	1.28	1015	0.76
AlSb	3.0	1323	1.60	1350	> 1.0
CdTe		1372	1.50	1235	0.9

^{a}Refractive index

Recent work by Wallace et al. [110] suggests that semi-transparency may also be a feature of some undoped (or very lightly doped) semiconductors. For semiconductors for which the band gap is greater than the energy corresponding to the peak in the black body spectrum of the hot crystal, the efficiency of internal radiative heat transfer and the optical depth will depend on the number of free carriers. If these are thermally generated (i.e. if the crystal is intrinsic) then there will exist some transition temperature below which the crystal becomes semi-transparent to its own internally generated thermal radiation. This manifests itself as an enhancement in the radial temperature gradient near the periphery of the crystal. These materials generally have a high refractive index and therefore a large Brewster angle so that almost all of the self-radiation incident on the crystal surface from within is reflected back into the crystal.

Scaling from experimental data obtained from eddy current monitoring of the temperature profile in a cooling crystal of silicon, Wallace et al. [110] have calculated a notional 'transparency temperature' $T(x)$ for a number of commercially important semiconductors. This is reproduced in Table 5.1. What is important is not the absolute value of $T(x)$ but its value relative to the melting point of the semiconductor. From Table 5.1 it can be seen that this normalised temperature $T(x)/T_M$ is 0.5 for silicon and > 0.7 for GaAs, InP, AlSb and CdTe. This result is of consequence to the evaluation of the thermal stresses in the cooling crystal (see Chap. 10).

Modelling the temperature distribution in a cooling semi-transparent crystal is very difficult because it becomes a non-local problem with boundary shape and reflectivity being critical factors. Useful approaches to this problem have been devised by Zalewski and Zmija [226] and by Kvapil et al. [227].

6 Mass Transport and Solute Segregation

6.1 Introduction

Most crystals required for commercial application need to be doped with a deliberately added solute in order to acquire the needed properties. (One significant exception to this is undoped semi-insulating gallium arsenide). The device specification will require that the doping level is precisely controlled and that it be uniform throughout the crystal boule both on a macroscopic and on a microscopic scale. Since the pulling process is essentially a batch process, segregation of solute at the interface results in a progressive increase in solute concentration in the crystal as the melt is consumed for solutes having a segregation coefficient (k) less than unity. The reverse is true for solutes with k greater than unity.

To overcome this problem of macrosegregation, a number of strategems are possible. Firstly, if the dopant has a segregation coefficient near to unity, it will be preferred. Such is the case for phosphorus, oxygen and boron in silicon. Failing this, one frequently accepts the fact of macrosegregation but uses only the first part of the grown crystal where the solute concentration is most uniform. More esoteric approaches which are not widely deployed include the use of an inner floating crucible (explained below), use of a time-varying magnetic field to change the effective distribution coefficient (see Sect. 7.6.3) or removal of the batch process limitation by arranging to steadily feed new charge material into the melt.

These options address the problem of the macroscopic uniformity in an axial direction. There remains the question of the radial uniformity of dopant across the crystal cross-section. This is controlled by the nature of the flow in the melt and a major benefit of the pulling technique is the control over the solute boundary layer adjacent to the interface which is provided by the crystal rotation as explained in Chap. 4. This generates what is known as a uniformly accessible boundary layer and ensures a radial uniformity which, in favorable cases, can be better than one percent.

The micro-uniformity of the dopant is determined by fluctuations in the process which can arise either from thermal or mechanical infelicities in the apparatus or from turbulent convection in the melt. This is considered in Sect. 6.4 below.

Because melt growth is a high temperature process which, by and large, is transport limited the segregation coefficient at the crystal-melt interface is, in general, near to its thermodynamic equilibrium value i.e. that value obtained by taking the ratio of the solidus to liquidus concentrations at the growth temperature from the phase diagram (Fig. 6.1). The principal exception to this is where a low index surface is tangential to a convex crystal-melt interface and where that surface exhibits a kinetic limitation to growth. In this case the interface lags behind the isotherm corresponding to the thermodynamic freezing temperature and a small faceted area of interface develops (Fig. 6.2). It has been well established [94] that segregation on such faceted surfaces occurs in a non-equilibrium manner and the segregation coefficient can differ quite dramatically from its equilibrium value. This is true both of semiconductor and oxide crystal growth. The most outstanding example is the segregation coefficient of tellurium

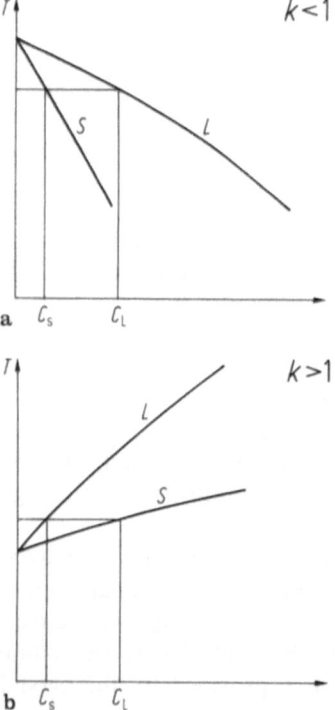

Fig. 6.1. Schematic phase diagram of a system with segregation coefficient k. **a** k < 1, **b** k > 1. Solidus-S; liquidus -L. Crystal and melt concentrations C_s and C_L respectively

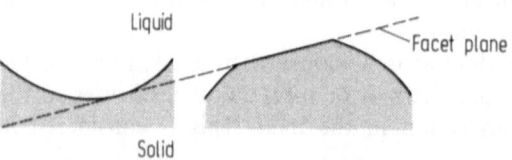

Fig. 6.2. Facet development (schematic). Facets can form only when the interface is convex-to-the-melt (*right hand diagram*). Facets do not form when the interface is concave-to-the-melt because adjacent regions of the crystal can always feed step growth

in indium antimonide first studied by Mullin and Hulme [95] who showed that the equilibrium segregation coefficient was approximately 0.5 whereas the segregation coefficient on (111) faceted surfaces was ≈ 2.0. i.e. the two segregation coefficients lay on opposite sides of unity, on and off the crystal facet. This effect, dubbed the facet effect, is potentially a source of major macroscopic inhomogeneity in the crystal as can be seen from Fig. 6.3.

Facets and a segregation facet effect are also seen in many oxide crystals. The lattice strain associated with such facets is readily observable by optical means (Fig. 6.4). Extensive studies of garnet crystals [96] has shown that the mean

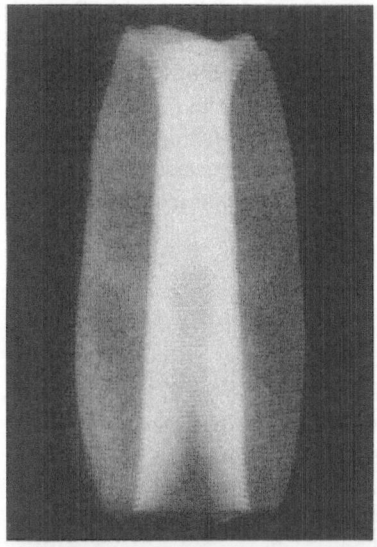

Fig. 6.3. Autoradiograph of a longtitudinal section of a ⟨111⟩ oriented InSb single crystal showing tellurium segregation at a {111} facet. Light regions are tellurium-rich corresponding to a higher segregation coefficient on the facet. The fine structure is the striation pattern due to slow crystal rotation. (Mullin [29]) (Mag × 1.7)

Fig. 6.4. Optically delineated strain pattern in a $Y_3Al_5O_{12}$ single crystal produced by the presence of three {211} facets on the growth interface. (Cockayne [96]) (Mag × 55)

lattice parameter is slightly greater in the faceted area and that this is most likely due to oxygen segregation caused by oxygen vacancy entrapment at the growth interface as growth steps sweep rapidly across the faceted interface.

What one measures experimentally is an effective segregation coefficient (k_{eff}) which is the ratio of the concentration in the crystal to that in the bulk of the melt. In transport limited conditions where there exists a boundary layer of solute concentration adjacent to the crystal-melt interface, this quantity will differ from the interface segregation coefficient (which is equal to the ratio of the concentration in the crystal to that in the melt at the interface). The degree of difference between these two quantities will depend on the nature and strength of the flow in the melt and on the crystal growth speed. Where this flow is dominated by natural convection, as discussed in Chap. 4, the relationship between these quantities can be obtained only from a full numerical simulation of the melt flow. However, it is common for the flow due to the rotating crystal to determine the solute boundary layer thickness in Czochralski growth and, in this case, a simple relationship exists between k and k_{eff} which was first obtained by Burton, Prim and Slichter [97]. This relationship is derived in the following section.

6.2 The Burton, Prim and Slichter Analysis

Mass transport is described by Fick's law of diffusion in its second order conservation form with coordinates transformed to move with the suction velocity produced by pulling, (v_p). The equation is:

$$dC/dt = D\nabla^2 C + v_p \, dC/dz \qquad\qquad 6.1$$

where D is the diffusion coefficient and C the solute concentration. In the steady state, and for a one-dimensional problem, we can employ the usual similarity transformation (Eq. 4.22) which, together with a non-dimensionalisation of the solute concentration:

$$\phi(\zeta) = (C(\zeta) - C_s)/(C(0) - C_s) \qquad\qquad 6.2$$

gives us the equation

$$\phi'' = Sc \, H(\zeta)\phi' \qquad\qquad 6.3$$

where $H(\zeta)$ is the Cochran solution and the prime indicates differentiation with respect to the parameter ζ. The relevant boundary conditions are, at $\zeta = 0$:

$$\phi = 1$$

and

$$[C(0) - C_s]v_p + DdC/dz = 0$$

Because the Schmidt number (Sc = v/D) is much greater than unity, the solute boundary layer is very thin compared to the momentum boundary layer so that, over the thickness of the solute boundary layer for which the concentration is changing, the flow field can be represented by the first term in the expansion for H in Eq. 4.31. With this approximation we can integrate Eq. 6.3 to obtain:

$$\phi(0) = 1 - J \tag{6.4}$$

where
$$J = [v_p(v/\Omega)^{1/2}/D] \int_0^\infty \exp\left[Sc - \int_0^x H(z)\ dz \right] dx \tag{6.5}$$

A more rigorous analysis has been performed by Riley and Sweet [257]. Noting that $C_s/C_0 = k$, we obtain for the effective segregation coefficient:

$$k_{eff} = k/[1 - (1 - k)\ J] \tag{6.6}$$

Burton, Prim and Slichter have related the parameter J to a boundary layer thickness, δ, by the relation:

$$v_p\delta/D = -\ln(1 - J) \tag{6.7}$$

However, this has been criticised by Wilson [98] and by Wheeler [99]. Wilson demonstrates that a more sensible definition is:

$$v_p\delta/D = J \tag{6.8}$$

These two definitions of course approach each other as J becomes very small compared to unity. Evaluating the integral in Eq. 6.5 gives:

$$\delta = 1.6Sc^{-1/3}(v/\Omega)^{1/2} \tag{6.9}$$

This gives us the familiar Burton, Prim and Slichter equation:

$$k_{eff} = k/\{k + (1 - k)\ \exp(-v_p\delta/D)\} \tag{6.10}$$

In the limit that δ goes to zero, k_{eff} becomes k i.e. when the boundary layer is completely obliterated by very strong convective flow then segregation approaches its equilibrium state. In the other limit, as the boundary layer thickness becomes infinite ($\delta = \infty$), k_{eff} approaches unity, i.e. the concentration of solute incorporated into the crystal is equal to that in the bulk of the melt.

6.3 Macrosegregation

This is described by the familiar Scheil equation. To give an insight into the conditions under which it is valid, a brief derivation follows:
   The assumptions made are:
1) that solute mixing in the melt is complete,
2) there is no diffusion in the solid,
3) a steady growth rate is employed,
4) the crystal-melt interface is planar and normal to the macroscopic direction of solidification,
5) the segregation coefficient k is constant.

Consider the normal freezing process exhibited schematically in Fig. 6.5. A crystal of unit circular cross-section is joined to a molten charge whose length at time t is $L - l(t)$ and whose freezing interface is moving at velocity v_p. The total length of charge plus crystal is L, $l(t)$ therefore being the crystal length. The crystallised material has a solute concentration $C_s(z)$, the liquid charge a concentration C_L. z is the coordinate in the direction of freezing. The rate of change of solute concentration in the liquid is:

$$d((L - l)C_L)/dt = -C_s v_p \qquad\qquad 6.11$$

where $C_s = kC_L$

If v_p is constant (i.e. independent of time), then the time dependence of the length of the liquid charge is given by:

$$l(t) = l_0 + v_p t \qquad\qquad 6.12$$

where l_0 is the solidified length at time $t = 0$. Equation 6.11 then becomes:

$$dC_L/\{(1 - k)v_p C_L\} = dt/\{L - l_0 - v_p t\} \qquad\qquad 6.13$$

where C_0 is the initial concentration in the liquid charge at time $t = 0$. Evaluation of this integral gives the solute distribution. Initially the whole charge is molten and $l_0 = 0$. Integrating Eq. 6.13 and assuming that the segregation coefficient k is constant, we obtain directly the Scheil equation:

$$C_s = kC_0(1 - g)^{k-1} \qquad\qquad 6.14$$

where $g = v_p t/L$ is the fraction solidified after time t.

One can measure the segregation coefficient of a solute by growing a crystal and then measuring the distribution of that solute down the crystal. Taking logarithms of Eq. 6.14:

$$\log(C_s/C_0) - \log k = (k - 1)\log(1 - g) \qquad\qquad 6.15$$

so that, from a plot of $\log(C_s/C_0)$ versus $\log(1 - g)$, one obtains an intercept of $\log k$ and a slope of $(k - 1)$.

However, we have assumed complete solute mixing in the melt. Whilst this will be approached under conditions of strong crystal rotation and/or of strong buoyancy-driven convection in the melt, in general the quantity measured will be an effective segregation coefficient in the spirit of Burton, Prim and Slichter as described above. For typical growth speeds employed for the growth of say gallium arsenide, effective segregation coefficients are perhaps up to twenty percent above their equilibrium values.

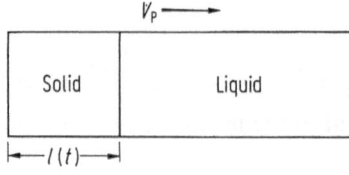

Fig. 6.5. Normal freeze configuration. Growth velocity V_p

A plot of the Scheil equation for different values of the segregation coefficient is shown in Fig. 6.6. It can be seen that solutes with k ≪ 1 have a distribution such that the concentration in the crystal doubles after half of the melt has been solidified and thereafter increases much more rapidly. Hence, even with the weak specification for control of solute concentration to within say a factor of two, one is limited for such solutes to using only the first half of the grown crystal. With solutes with k close to unity, the macrosegregation is much weaker. For solutes with k greater than unity, the solute is enriched at the first end of the crystal to solidify.

The macrosegregation can be diminished somewhat by having a variable k_{eff} so that, for k < 1, one initiates growth under conditions of poor stirring so that $k_{eff} \gg k$, increasing the stirring and thereby decreasing k_{eff} as growth proceeds. As previously indicated, for conducting melts one can also change k_{eff} by the application of a magnetic field (see Chap. 7).

Yet another approach is to float an inner crucible in the melt, weighted so that it is partially submerged. Growth is initiated on the seed from the melt in the inner crucible. That melt is continuously replenished from the melt in the outer (main) crucible via a narrow connecting orifice in the base of the inner (floating) crucible. (Leverton [228] and Goorissen [229]). Provided that there is no diffusive or convective mixing between the inner and outer crucibles (but only Poiseuille flow through the narrow connecting orifice) then the inner melt is continuously fed by melt of *constant* composition so that, after an initial transient, the solute composition of the crystal assumes a steady value equal to that of the melt in the outer crucible.

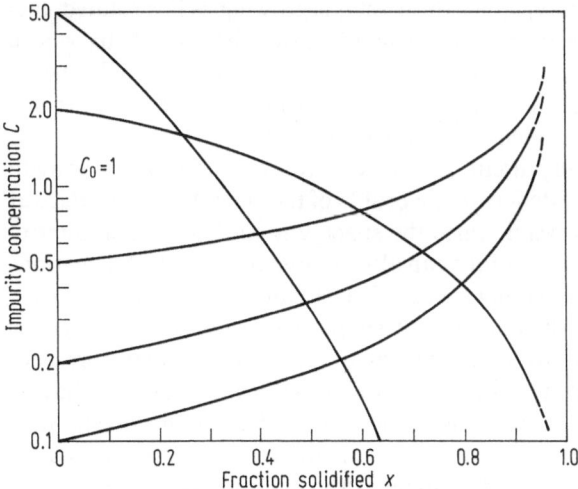

Fig. 6.6. Normal freeze solute distribution (Scheil equation) shown for solutes having segregation coefficient k = 0.1, 0.2, 0.5, 2 and 5

Whilst elegant in concept, the technique inevitably poses difficult problems of mechanical and thermal engineering. Some design considerations have been reviewed by Brice [230].

6.4 Microsegregation

There are essentially two major types of microsegregation in melt-grown crystals. The first of these is due to a fluctuation in the normal growth rate at an essentially planar crystal-melt interface. The second is the result of a morphological instability in the shape of the crystal-melt interface on a microscopic scale. Consideration of this latter phenomenon is deferred until Chap. 9.

The microsegregation which arises from a fluctuation in the normal growth rate gives rise to a pattern of solute distribution down the crystal which is banded. Such a distribution is commonly referred to as solute striations. There are a variety of potential causes of such a distribution. Firstly, in the case of Czochralski growth, the rotation of the crystal can be a potent cause of dopant striations particularly if the growth equipment is poorly designed so that it lacks thermal symmetry. A given point on the crystal-melt interface is carried on a circular path through the melt about the crystal axis. If this axis is not also the axis of thermal symmetry of a purely radial temperature field then, as the crystal rotates, the considered point on the interface will be taken through a periodically varying temperature field. This will cause the rate of growth at that point to oscillate. As it oscillates, so it will modulate the magnitude of the solute boundary layer ahead of it in the melt (for all solutes having a segregation coefficient not identical to unity). Since the crystal is growing as it is rotated, this pattern of modulated solute distribution will be in the form of a helix whose pitch is equal to the amount of crystal pulled per crystal revolution. Such a helix, sectioned vertically in any plane away from the axis of the helix, will reveal a banded distribution of solute such as is shown in Fig. 6.7.

These striations are a very useful diagnostic tool for inferring the growth sequence post hoc, since they delineate the profile of the crystal-melt interface at successive rotations of the crystal. Since the spacing is equal to the amount of crystal grown per revolution, striations can also be used to infer local deviations in the growth speed from the pulling speed. Also, as can be seen from Fig. 6.8, if there is some discontinuity in the smooth shape of the surface, such as the development of a facet, then this too is delineated. Figure 6.8 shows the presence of a near-central (111) facet which meanders as growth proceeds, and in one section just beneath the shoulder, disappears altogether. If the crystal-melt interface is curved, then this helix of solute enhancement will be 'dished'. A section through the crystal taken orthogonal to the pulling axis will then reveal solute striations in the form of a spiral (Fig. 6.9a). Truncation of the curved interface by a crystal facet is evident in the figure.

Fig. 6.7. Rotational striations in an arsenic-doped germanium crystal revealed by etching a longtitudinal section (Mag × 80)

These, so-called rotational striations [100], are readily seen in the Czochralski growth of relatively low-melting point semiconductors grown from moderately sized melts. However, as the melt size is increased and, with high melting-point oxides at essentially all sizes, these striations are replaced by others which in general, are not regularly spaced and do not represent the amount of crystal grown per revolution. (Fig. 6.9b). The latter owe their origin to turbulent convection in the melt. The occurence of such turbulence was considered in Chap. 4. There are essentially two mechanisms which cause this class of striation. Firstly, the fluctuating (turbulent) flow produces a fluctuating temperature at a point in the melt (Fig. 6.10). This fluctuating temperature propagates towards the crystal-melt interface where it is largely absorbed as latent heat and produces a fluctuating growth rate. This in turn, by the mechanism described above, gives rise to a fluctuating solute concentration. The second mechanism arises from the direct influence of the flow on the solute field. Thus a fluctuation in the momentum boundary layer due to fluctuations in a shear flow across the interface will in turn modulate the solute boundary layer. This modulation will tend to be rather weak because the solute boundary layer is embedded within the momentum boundary layer. Nonetheless, for oxide melts at least and with relatively slow rotation rate, the strength of the turbulence in the melt is such that this is believed to be an important mechanism.

The response of the interface to temperature fluctuations propagated from the melt has been analysed using linear perturbation theory [102]. This shows that there is a phase-lag between an oscillation in the temperature field and the resulting oscillation in the concentration field in the crystal. In elegant work, Witt and co-workers [100] have demonstrated this experimentally in pulled crystals of gallium doped germanium as shown in Fig. 6.11.

Fig. 6.8. Longtitudinal section of a doped germanium crystal grown 7° off the ⟨111⟩ axis. Striations revealed by pulse plating. Note the {111} facet. (Dikhoff [238]) (Mag × 1.8)

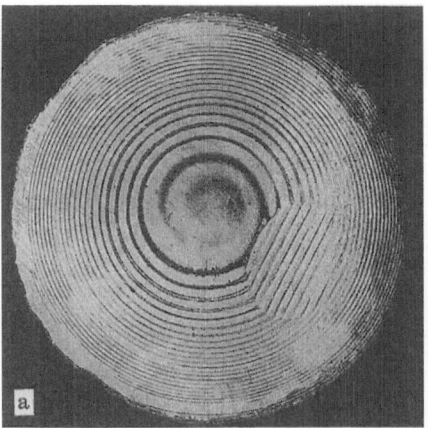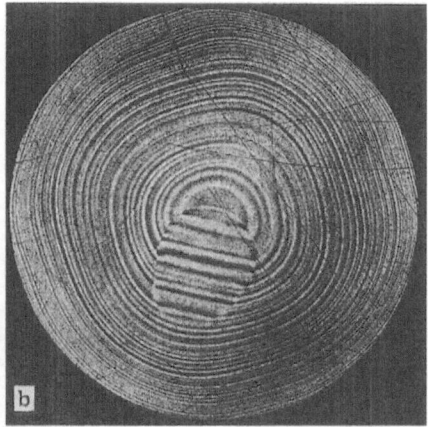

Fig. 6.9. a Transverse section of the crystal shown in Fig. 6.8. Rotational striations form a single start spiral the arm of which is distorted by the presence of the {111} facet. **b** Similar configuration for a $Ba_2 NaNb_5 O_{15}$ crystal with a {001} facet revealed by etching. Note that the striae in **b** are concentric having been produced by temperature fluctuations in the melt rather than by rotation. (Hurle and Cockayne [240]) (Mag × 3)

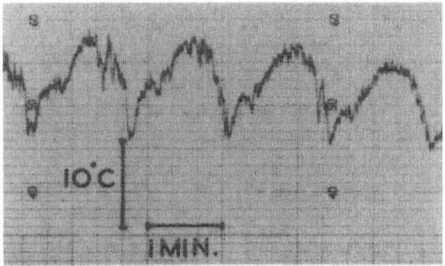

Fig. 6.10. Temperature oscillations and flu-ctutations in a $CaWo_4$ melt. (Cockayne and Gates [239])

This fluctuation alters the spatially-averaged segregation coefficient as shown by Hurle and Jakeman [101]. Put simply, if the temperature in the melt oscillates sinusoidally causing a sinusoidal oscillation in the interface velocity, then during the half cycle for which the growth velocity is above its mean value, and with it the solute concentration at the interface is above its mean value, more crystal is grown than during the other half cycle. Consequently, for a segregation coefficient less than unity, the effective segregation coefficient is increased by a melt temperature fluctuation. The maximum value of this increase in the absence of melting-back of already grown crystal, is a factor of two. If the segregation coefficient is greater than unity then the spatially-averaged effective segregation coefficient moves downward toward unity. This has been demonstrated experimentally by Bartel and Eichler [48] in electron beam zone melting of tungsten-doped molybdenum.

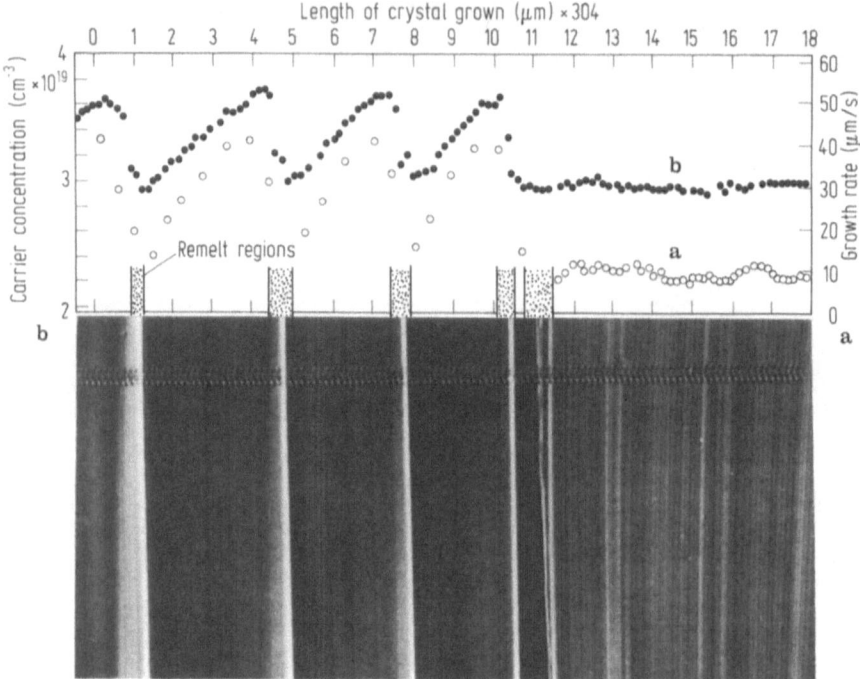

Fig. 6.11. Longtitudinal section of a Ga-doped Ge crystal showing striations produced by crystal rotation. Between these are much finer 'rate' striations produced by passing a modulated electric current through the crystal. Also shown is the carrier concentration profile obtained from spreading resistance measurements and the instantaneous growth rate obtained from the spacing of the rate striations. The growth rate follows the thermal field; note the phase lag of the dopant concentration peak with respect to the growth velocity. (Witt et al. [100])

If the amplitude of the temperature fluctuation is so great as to cause melting back of part of the already grown crystal, then linear perturbation theory [102], [103] becomes invalid and the changes in the effective segregation coefficient are more complex [104], [105]. In particular Wilson [106] has studied the situation explicitly relating to crystal pulling, i.e. at a rotating disc. A complete analysis including the case of melting-back has been given by Wheeler [107].

The concentration gradient associated with dopant striations introduces stresses into the cooling crystal. If these exceed the critical resolved shear stress of the crystal then dislocations are introduced to relieve the elastic strain (Fig. 6.12). This topic is treated in Chap. 10.

Finally we give consideration to prescriptions for minimising the occurrence of solute striations since this is a usual objective for the commercial growth of materials for device application.

The first approach is clearly to improve the mechanical and thermal design of the puller so as to minimise the amplitude of the rotational striations. Having done this the avoidance of time-periodic convective flow is central to the

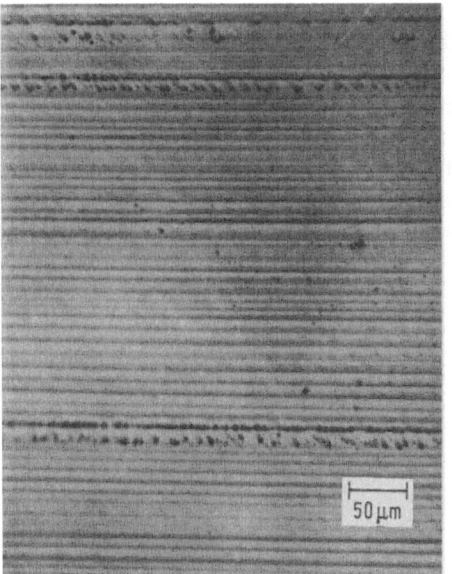

Fig. 6.12. Longtitudinal section of a Ge-Si alloy single crystal showing dislocations (revealed as etch pits) formed on striations to relieve the elastic stress caused by the lattice parameter difference between Ge and Si. (Goss et al. [241])

avoidance of striae. The use of a magnetic field is useful to this end. This is considered in detail in the next chapter. However, it only applies to conducting melts and provides no answer to the problem of avoiding striae in oxide crystals. An esoteric approach would be to grow crystals in a gravity-free environment, (i.e. in space) but a moment's consideration will show that there are inherent problems in applying the Czochralski technique in space since it depends on gravity to maintain the configuration of the melt contained within a crucible.

Minimising the melt turbulence requires that one minimises the Grashof number. To do this one has to reduce the temperature differences across the melt, i.e. make the pulling environment more isothermal and reduce the physical dimensions of the melt volume. Making the puller more isothermal brings difficulties in the control of the crystal diameter. Reducing the effective melt dimension can be achieved by introducing baffles [108] or a secondary inner crucible into the melt [109]. To minimise the amplitude of the concentration fluctuation in the striation, it is most important to avoid melt-back. Melt-back is least prone to occur if the growth rate is relatively fast and the amplitude of the temperature oscillation small. Again this is not readily achievable in oxide growth where very low growth speeds are required in order to maintain morphological stability of the interface (see Chap. 9).

The other parameter that can be considered is the frequency spectrum of the temperature fluctuations. Calculations indicate that there is a very strong response of the interface under specific conditions to rather low temporal frequencies and Birman [103] has demonstrated that this response is resonant. A detailed understanding of the flow mechanisms and of the temporal frequencies generated might permit some more refined control of the process. Of course different applications for the crystal will require different specifications of homogeneity and the spatial scale for the striations is dictated not only by the temporal frequency of oscillation, but also by the speed of crystallisation. If $\Omega/v_p > 10^4$ (where Ω is the temporal frequency of temperature oscillation) then the striae will be less than 2π microns apart and solid state diffusion will tend to smooth them out. If $\Omega/v_p < 1$, the striae are more than 2π cm apart and therefore approaching the dimensions of a typical crystal.

7 The Use of a Magnetic Field

7.1 Motivation

The idea for the use of a magnetic field to damp melt turbulence and thereby improve microscopic homogeneity of the crystal was introduced in 1966 independently by Utech and Flemings [112] and by Chedzey and Hurle [113]. However its application to crystal pulling did not come until much later. In the late 1970s, workers with large silicon pullers observed systematic anomalies in oxygen content in crystals grown on apparently similar pullers. Specifically it was found that differing oxygen contents could be obtained depending on the sense of the crucible rotation, always with the seed rotating counter to the crucible [114]. Eventually it was discovered that this was due to the rotating magnetic field generated by the three-phase heater elements.

In 1981 workers at Sony in Japan [115] developed a growth process using a static transverse magnetic field generated by an external electromagnet. Since that time there has been considerable research into the effect of externally applied magnetic fields on the growth of silicon and of the III–V compound semiconductors, gallium arsenide and indium phosphide (see [116] for a review).

Motivation for the use of an applied magnetic field has widened subsequently. In addition to damping out the turbulence and thereby removing the dopant striations, the field can be used to control the growth conditions at various stages in the growth process and this allows the production of silicon crystals with oxygen contents spanning more than an order of magnitude. This latter is its most important potential application. In the case of III/V compound semiconductor growth, not only is there a need to damp out microscopic fluctuations in dopant concentration but, because the melt from which the crystal is being grown is in general not at its congruent composition, a fluctuating growth rate also produces fluctuations in crystal stoichiometry which are to be avoided. In the case of gallium arsenide it has been established that the concentration of a very important native defect (EL2) which forms a deep level near mid-band gap and which controls the electrical behaviour of undoped semi-insulating material, is dependent on the crystal stoichiometry [117]. Finally, there is the possibility of using a magnetic field as a secondary control variable in automatic diameter control.

7.2 Interaction of a Magnetic Field with Melt Flow

The basic mechanism for the interaction between a magnetic field and a molten metal or semi-conductor is the interaction between it and electrical current induced in the conducting melt by its motion in the DC magnetic field. This current flows such as to damp the velocity field.

The Lorenz force F acting to damp the motion is then given by:

$$F = J \times B \qquad\qquad 7.1$$

where J is the induced current and B the magnetic induction. The dissipation is taken account of through an extended form of Ohm's law:

$$J = \sigma(E + u \times B) \qquad\qquad 7.2$$

where u is the flow velocity, σ the electrical conductivity and E the electric field The electrical parameters J and E can be eliminated using Maxwell's equations to obtain the following equation for the magnetic induction:

$$\dot{B} = \nabla \times (u \times B) - (\mu_0 \sigma)^{-1} \nabla \times (\nabla \times B) \qquad\qquad 7.3$$

where μ_0 is the free-space permeability. This is coupled to the Navier Stokes equation to which a Lorentz force term has been added.

Two material specific magnetic parameters are of importance. These are the magnetic Reynolds and magnetic Prandtl numbers. The magnetic Reynolds number is:

$$Re_m = \sigma \mu_0 Ud$$

(U is a characteristic flow velocity and d a typical system dimension). This is a measure of the relative magnitudes of the convection and diffusion terms on the RHS of Eq. 7.3. In all crystal growth situations involving metallic melts, $Re_m \ll 1$ and the diffusion term dominates. This means that the flow distorts the field lines by only a very small amount.

The magnetic Prandtl number is:

$$Pr_m = v/v_m$$

where $v_m = (\sigma \mu_0)^{-1}$ can be thought of as the diffusion coefficient for the magnetic field. This is very large for liquid metals so that $Pr_m \ll 1$ (typically of the order of 10^{-6}).

The small magnitudes of Re_m and Pr_m permit significant simplification of the Navier Stokes equations. Baumgartl and Müller [259] have recently shown however that it is not permissible to totally neglect the distortion of the imposed field by the flow. The interested reader should refer to their paper to obtain the full equations necessary to describe the magneto-hydrodynamic flow.

The geometry of a Czochralski melt is sufficiently complicated that a detailed calculation of the effect of an applied magnetic field on the motion requires extensive finite element or finite difference computation. Some general observations can be made however. Firstly, since B appears both in the expression for the induced current (Eq. 7.2) and for the damping force (Eq. 7.1), most of the interactions will scale roughly with the square of the field. Secondly, the damping force is the vector product of magnetic field and induced current. This implies that there can be no direct damping of flow parallel to the field lines; it is only where the flow cuts lines of flux that a Lorenz force occurs. It must however be noted that, as all the melt is constrained within a crucible and is incompressible, most flows will at some point be damped and exert back pressure on the undamped regions. An obvious exception is solid-body rotation with the magnetic field parallel to the rotation axis.

The concept of "magnetic viscosity" is an appealing one. This expresses the fact that the effect of the field is in general similar to the effect of increasing the viscosity of the melt. However, care must be taken not to push this analogy too far as the magnetic interaction is anisotropic as we have just seen whilst the viscosity of a Newtonian liquid is isotropic.

7.3 Generation of the Magnetic Field

Experience to date suggests that useful fields lie in the range of 500–5000 gauss or perhaps somewhat higher. These have to be generated in a large volume. For an axial magnetic field, the bore of the solenoid must be large enough to contain the crucible heater and any necessary heat shields. For, say, 15 cm diameter silicon this could require a bore diameter approaching one metre. Such fields can be generated either by resistive or by superconducting magnets.

For resistive electromagnets the resistivity of the winding is the key parameter determining the dissipation and the attainable maximum field. For example, the power dissipation in a 0.5 m bore copper coiled magnet generating a field of 2 kg is the order of 100 kW. This is a comparable power to that required for the heater of a puller to which this might be fitted and used for the growth of silicon. Clearly considerations of power dissipation favour the use of a superconducting magnet but this requires more maintenance and is less amenable to modification. Resistive magnets on the other hand are simpler to operate and easier to fabricate and modify. Although the at-field running cost is lower for superconducting designs, this is largely negated by the liquid helium loss whilst not in use. Both types of magnet have a very considerable stored energy which generates large stresses on the windings. Cost analysis shows that above some critical size of puller, for a given required field, superconducting designs win out.

Coupling the magnet to the puller is not straightforward since the field will couple with the current in the resistive puller heater element. Many heaters are operated with a phase-angle-controlled single phase AC supply. To prevent unacceptable vibration of the heater in the field this must be replaced by a full-wave bridge-rectified supply. Even with a DC-fed heater there are considerable forces on it which may require re-engineering of the heater element.

7.4 Magnetic Field Configurations

Several different geometries of applied magnetic field have been reported in the literature. Early work utilised conventional electromagnets with pole pieces to give a transverse field. (i.e. transverse to the pulling axis). Later, solenoids were fitted around the puller to give an axial magnetic field. Deficiencies in both of these configurations became apparent which led to the development of a configured magnetic field in which the field lines are splayed about an axis coincident with the puller axis.

The simplest configuration of the transverse field is the conventional electromagnet (Fig. 7.1). More sophisticated designs have utilised a soft-iron yolk in the form of a toroid placed around the puller. At opposite ends of the diameter two pole pieces are fitted, around which are placed electromagnetic coils. The magnetic circuit is then essentially that due to two "C" shaped magnets clamped symmetrically around the diametral plane. The principal limitation of such a field is that it destroys the axial symmetry of the pulling process and hence exacerbates rotational striations [118].

With an axial field, the melt is positioned within the centre of a simple solenoid mounted with its axis coincident with the pulling axis (Fig. 7.2). More than a single coil can be utilised to trim the field uniformity [119]. Such a field maintains the axial symmetry but, as we shall see, gives a very poor radial uniformity of dopant.

The concept of a configured field [24], [25] was developed to tailor the field configuration to the flows in the melt so as to damp the harmful flows whilst retaining the beneficial flow patterns. One such configured field which has received extensive experimental study is the cusped field (Fig. 7.3) formed with a pair of Helmholtz coils operated in opposed-current mode.

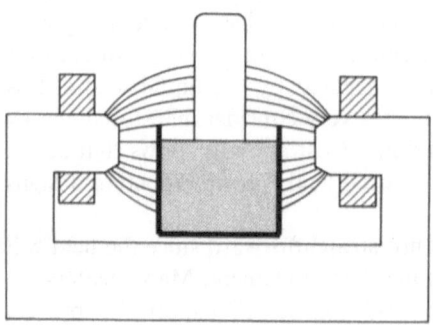

Fig. 7.1. Schematic representation of the transverse magnetic field configuration

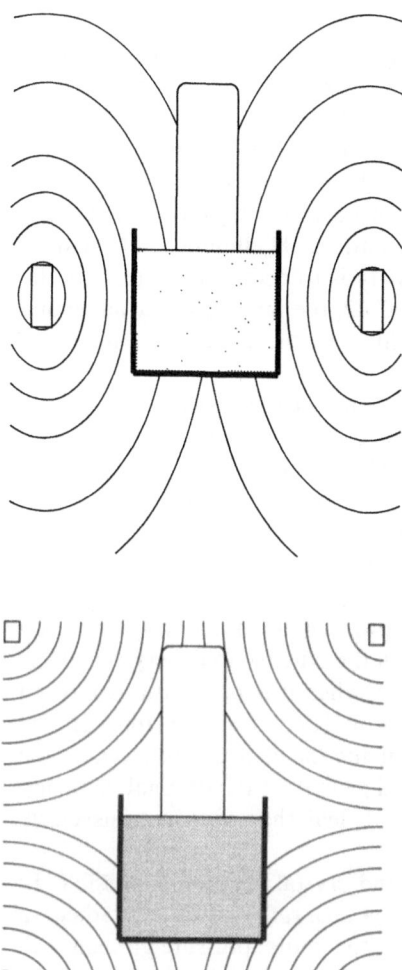

Fig. 7.2. Schematic representation of the axial magnetic field configuration

Fig. 7.3. Schematic representation of the cusped magnetic field configuration

The most extensive work to date has been on the axial magnetic field and in the next section we consider the theory of flow and segregation in the presence of such a field.

7.5 Flow and Segregation in a Magnetic Field

7.5.1 Flow and Temperature Fields

We have already seen how complicated the flow is in a Czochralski melt. Whilst the presence of a magnetic field might be thought to complicate it even further, there is one sense in which it makes the problem easier. Since the field damps the flows, it also reduces the velocity gradients in the various shear layers Numerical simulation of the flow pattern is therefore rendered somewhat easier since, if the shear layers are thicker, then a coarser grid mesh can be used to resolve them. The computational cost of simulating the magneto-hydrodynamic flow can therefore be less than that for the purely hydrodynamic problem. A number of such simulations have been carried out (see [116] for a review). Because of the large number of parameters involved, it is difficult to obtain much physical insight just from numerical simulation. Hjellming and Walker [120–122] have therefore used asymptotic analysis to deduce the nature of the boundary layers formed. As indicated in Sect. 7.2, the magneto-hydrodynamic equations to be solved are considerably simplified by the fact that the magnetic Prandtl number (Pr_m) is very small.

Langlois and Lee [123] have shown that joule heating effects also can be neglected. Hjellming and Walker [124] have shown, somewhat surprisingly, that the finite (i.e. non-zero) conductivity of a semiconducting crystal can profoundly effect the azimuthal motion and this has been confirmed in numerical simulation by Langlois [125]. However, since it is only the azimuthal motion which is affected, this does not have a profound influence on the segregation of solute.

Because it damps the flow, the application of an axial magnetic field significantly reduces the heat transfer rate from the crucible to the crystal. Since, in order to grow a crystal of fixed dimensions, when it is being drawn into an environment having fixed thermal properties, a fixed heat flux through the crystal-melt interface is required, then, on applying the field, one has to raise the crucible wall temperature in order to maintain growth at the original diameter. This effect is more pronounced with an axial field than with a transversely applied one [126].

Some insight into the effects of crucible and crystal rotation is possible. In the absence of a crystal, crucible rotation will set up a radial pressure gradient in the melt. If the melt is rotating as a solid body, this radial pressure gradient will be the same at all heights and so there will be no driven flow (i.e. the flow will be geostrophic). If we now consider the effect of advected liquid rising up the crucible wall and flowing inwards, then this liquid will carry excess angular momentum and so will start to increase in angular velocity as it moves inwards. This will in turn, increase the radial pressure gradient at the top which will oppose the advected flow. It can be seen therefore that, in the absence of a crystal, crucible rotation decreases the convection. If we now consider the effect of a crystal only weakly coupled to the system, again crystal rotation will have little effect. If however, the crystal influences the flow field to a significant depth, then there will be a significant imbalance between the radial pressure gradient at the top and bottom of the melt leading to a strong forced convection. Now it is found both experimentally and by analytic and numerical modelling that a magnetic field results in an increased coupling between the crystal and the melt. We can therefore conclude that one important effect of a magnetic field will be to increase the significance of crystal rotation in promoting flows in the bulk of the melt.

An understanding of the bulk flow and the manner in which it is affected by a magnetic field is necessary for an understanding of the thermal stability of the system and for a consideration of non-conservative doping behaviour where there are sources and/or sinks of dopant in the system. The most important example of this is the uptake of oxygen in Czochralski silicon (see Sect. 7.6.1 below). However, in most cases, the doping is conservative; the dopant concentration is uniform in a well mixed melt and varies only adjacent to the crystal-melt interface which acts as a differential sink for dopant and solvent. In this event only the flow field adjacent to the crystal-melt interface is of importance and an analytic approach using rotating disc theory is rewarding.

The modification to the Cochran flow derived in Chap. 4 to take account of the magnetic field involves a characteristic non-dimensional parameter N where:

$$N = B_0^2 \sigma_c / \rho \Omega \qquad\qquad 7.2$$

is a magnetic interaction parameter and B_0 is the intensity of a uniform axial magnetic field. The crystal is assumed to be electrically insulating. Employing the von Karman similarity transformations, Kakutani [127] has obtained an expression for the axial flow velocity which, in the high magnetic field limit, has the following form:

$$w(z) = -(v\Omega)^{1/2} N^{-3/2} [1 - 2 \exp(-N^{1/2}z) + \exp(-2N^{1/2}z)]/3 \qquad 7.3$$

In a weak axial field, the expression for the axial component of the flow is more complex. However, the overall form of the flow is qualitatively similar to the high field limit but with some small positive value for the parameter N in the case of zero field [128]. The magnitude of this modification to N can be obtained by matching to the Cochran solution to give:

$$N' = N + 0.427 \qquad\qquad 7.4$$

If this quantity is used to replace the parameter N in Eq. 7.3 then we have an adequate representation of the flow field for all values of the axial magnetic field.

The parameter N can be thought of as being related to the concept of magnetic viscosity mentioned above so that the total viscosity is proportional to N', the additive term (0.427) being a representation of the contribution of the kinematic viscosity to the term. One sees that the thickness of the momentum boundary layer (which is proportional to $(N')^{-1/2}$) is decreased by applying a magnetic field. In the next section we show how this affects the effective segregation coefficient of a solute.

7.5.2 Segregation

Substituting this newly obtained value for the axial component of the flow into the diffusion equation (see Chap. 6) and integrating subject to the same boundary conditions gives us an expression for the effective segregation coefficient;

$$k_{eff} = k/\{1 - (1 - k)J\} \qquad\qquad 7.5$$

where:

$$J = \int_0^{\infty} \exp\left\{ -v_p z/D + \int_0^{z^1} w(z)/w(0) \cdot (v_p/D) \, dz' \right\} dz \qquad 7.6$$

Note that this reduces to the Burton, Prim and Slichter relationship in the limit $N = 0$.

The problem is characterised by two dimensionless numbers:

$$\Theta = N' Sc^{-1/2} \qquad 7.7$$

and

$$\eta = v_p/D \cdot (v/\Omega)^{1/2} \cdot Sc^{-1/4} \qquad 7.8$$

η is a measure of the combined effects of suction due to crystallisation and rotation on the flow and Θ is a measure of the magnetic field strength. A plot of J as a function of η for several values of Θ is shown in Fig. 7.4. Increasing the growth velocity (increasing η) at fixed magnetic field (i.e. fixed Θ) gives an increase in J which causes k_{eff} to move towards unity.

The effect of the axial field can be perceived as follows. As the crystal rotates it rotates the liquid adjacent to it. This in turn creates a centrifugal pumping action and liquid is pumped radially outward and draws in fresh liquid axially towards the interface. As the crystal grows and dopant is rejected at the interface ($k_0 < 1$), the pumping action washes the dopant away from the interface. When an axial magnetic field is applied the field damps that radial flow which in turn suppresses the axial flow. The dopant now starts to accumulate at the interface and the composition of the solid and therefore k_{eff} increases. In zero magnetic

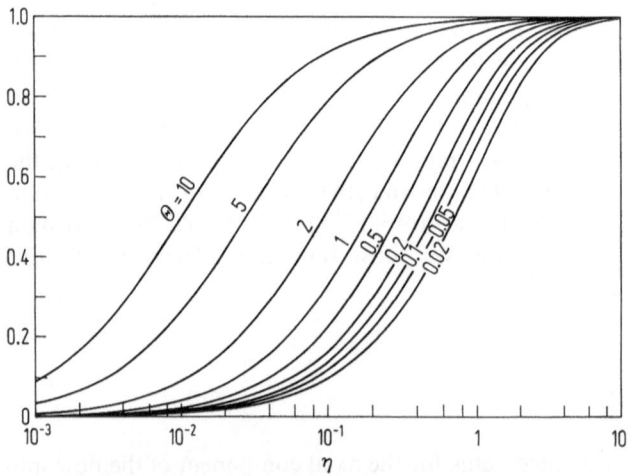

Fig. 7.4. Plot of segregation integral (J) – see text – versus normalised growth velocity (η) for several values of the magnetic interaction parameter (Θ). (Hurle and Series [128])

field the velocity of the axially-pumped flow usually dominates over the effect of crystal growth (i.e. of suction at the crystal/melt interface). However for axial magnetic fields of around 2000 gauss or greater the situation is reversed and the suction velocity due to the growth of the crystal makes the dominant contribution to the axial velocity term [129]. The parameter J corresponds to the normalised extent of the diffusion profile; small J means that the diffusion profile is confined close to the crystal and falls within the region of melt controlled by the rotating crystal. However, as J approaches unity, the diffusion profile extends into the region where the flow field is modified by the far field behaviour of the melt.

Because the above theory is based on a one-dimensional similarity transformation, the analysis necessarily predicts a uniform radial distribution of the dopant. We cannot therefore use it to calculate any radial variation but we can extract from it the condition for which radial uniformity can be expected. This condition is essentially that the solute boundary layer lies within a momentum boundary layer controlled by the crystal. The extent of the diffusion profile is given by Wilson [98]:

$$L_d = JD/v_p \qquad 7.9$$

The thickness of the momentum boundary layer is given by:

$$L_m = (v/\Omega N)^{1/2} \qquad 7.10$$

Thus good radial uniformity is to be expected only if $L_d < L_m$. Expressed mathematically this condition for good radial uniformity is then;

$$\int_0^\infty \exp\{ -\eta\Theta^{-1/2}z - [z + 2\exp(-z) - \tfrac{1}{2}\exp(-2z) \\ - \tfrac{3}{2}]/3\Theta^2\}dz < 1 \qquad 7.11$$

Analytical solution of this inequality is difficult; however the relationship can be evaluated numerically. In Fig. 7.5 a plot of η versus Θ is shown and on it is marked the regime in which the inequality above is not satisfied. There is also a lower limit to Θ for any given semiconductor (corresponding to N = 0) which depends on the value of the Schmidt number. This is marked on the graph for the case of silicon. Only growth in the unshaded region is expected to yield crystals with acceptable uniformity. To translate this graph into experimentally observable units it is necessary to specify either a growth rate or a rotation rate. As it is generally possible to vary crystal rotation over a much wider range than growth rate, the graph is replotted assuming a growth rate of 1 mm/min which is fairly typical for silicon. We can now plot, on the graph, contours of constant magnetic field (Fig. 7.6). From this we can see that for zero field, growth will always occur in a regime of good radial uniformity. However the application of quite a modest field of only 500 gauss requires a crystal rotation rate in excess of 30 rpm to achieve acceptable radial uniformity. To achieve an acceptable uniformity with a 2000 gauss field would require a crystal rotation rate in excess of 700 rpm which is not practicable. We therefore conclude and experiment

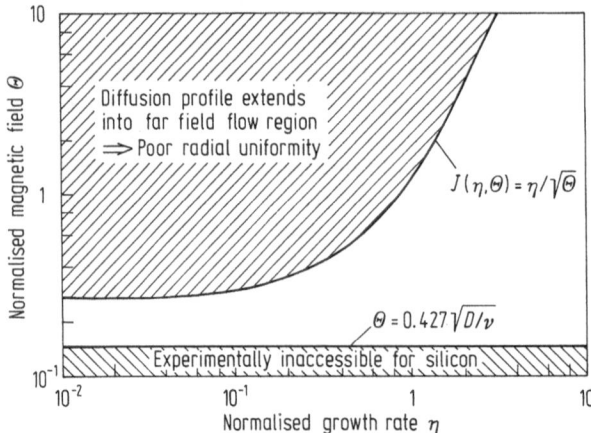

Fig. 7.5. Magnetic interaction parameter (Θ) versus normalised growth rate (η). *Shaded region* to upper left defines regime where radial uniformity is degraded. (see text). *Line* labelled $\Theta = 0.427$ $(D/v)^{1/2}$ corresponds to zero magnetic field. (Series [135])

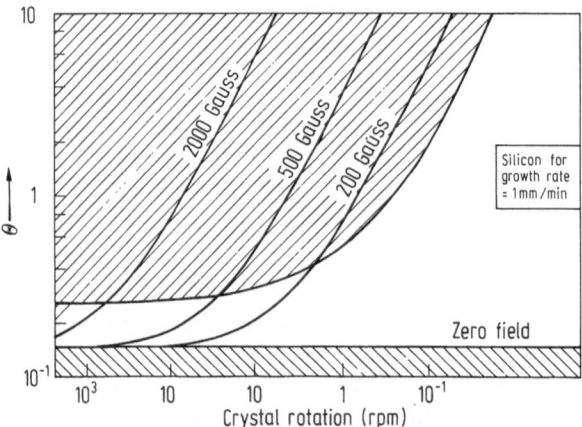

Fig. 7.6. Figure 7.5 replotted as a function of crystal rotation rate for a growth rate of 1 mm min^{-1} with parameter values appropriate to silicon. Contours for applied axial magnetic fields of 200, 500 and 2000 gauss are shown. Note that radial uniformity can be preserved at high magnetic field only by employing unrealistically high crystal rotation rates

confirms that an axial magnetic field, whilst damping the turbulence and removing micro-inhomogeneity, gives rise to a very poor radial uniformity.

We next consider a transverse field. Flow at a rotating disc in an infinite fluid is unaffected by a transverse magnetic field (provided that the electrical conductivity of the disc is negligible [130]). Hence the segregation of conservative dopants should not be affected by such a field provided that the "infinite disc" approximation is a relevant one. A more rigorous analysis taking account of the finite size of the crucible and the three-dimensional effects which this introduces has been carried out by Williams et al. [131]. To first order the infinite disc approximation is valid. What the transverse magnetic field does do is dramatically improve the microscale dopant uniformity by damping out the turbulence but, on the other hand, it can leave pronounced rotational striations which are caused by the thermal asymmetry which is introduced by the transverse field. These qualitative predictions have been experimentally verified by Ravishankar et al. [132] for phosphorus in silicon up to a field strength of 1500 gauss.

However, the uptake of oxygen in the presence of a transverse field is much more complicated on account of the change in the dynamic processes associated with dissolution of oxygen at the crucible wall. In general, the effect of the magnetic field is to inhibit flow in the momentum boundary layer, resulting in a thicker solute boundary layer which, in turn, can have the effect of reducing the solute flux away from the wall.

Clearly then, both axial and transverse magnetic fields have serious limitations. Both achieve the desirable result of damping out turbulence but they have different effects on solute segregation. The breaking of the geometrical symmetry of the pulling process, which results from application of a transverse magnetic field, accentuates rotational dopant striations. On the other hand, the use of an axial field which does maintain the symmetry, suffers the disadvantage that, as the magnetic field increases, the radial uniformity of doping is progressively degraded. This becomes serious at just the values of magnetic field needed to influence the oxygen uptake in silicon.

The solution to this problem was devised independently by Series [24], and by Hirata and Hoshikawa [25]. These workers proposed the use of a field configured so that, in the plane of the crystal-melt interface, it was purely radial whereas deep in the bulk of the melt it was purely axial. This field configuration was obtained by the use of a pair of Helmholtz coils operated in the opposed current mode. The diameter and separation of the coils can be chosen to maximise the radial component of field over the vertical crucible wall. In addition it is possible to arrange a large vertical component of field over the base of the crucible. Although the prime objective of this configuration is to reduce the oxygen content of the silicon crystal, a large fraction of the melt volume will be subject to magnetic damping so that there will be a reduction in the turbulence and concommitantly in the doping striations.

7.6 Applications

7.6.1 Silicon

Problems of oxygen control are unique to silicon, the oxygen content of the crystal playing a major role in determining the yield and performance of advanced integrated circuits [133]. The oxygen arises unavoidably from reaction between the molten silicon and the silica crucible wall. Most of this oxygen evaporates as silicon monoxide at the free-melt surface as it is advected around the crucible. A relatively small fraction (at most a few percent) of the oxygen becomes incorporated in the crystal. Control of the flow field within the melt can have a major impact on the exact amount of oxygen incorporated into the crystal. In addition to buoyancy-driven convection, Marangoni flow at the melt surface appears to have a large effect on the rate of evaporation of silicon monoxide. By contrast, the conditions under the rotating crystal appear less important since the most reliable experimental evidence to date suggests that the segregation coefficient of oxygen is close to unity [134] implying that there is no significant oxygen boundary layer ahead of the growing crystal. The pattern of oxygen incorporation is a convoluted map of the time variation of the oxygen distribution within the melt. This contrasts strongly with the case of normal dopants where their concentration in the bulk of the melt is essentially uniform and micro and radial segregation patterns in the crystal result from disturbances in the boundary layer adjacent to the crystal-melt interface.

Published experimental results of the use of the several field configurations on the oxygen uptake reveal this dependence on the bulk flow. Hoshi and co-workers [115] were the first to report on silicon crystal growth under an applied transverse magnetic field. They demonstrated that application of the field damped the temperature fluctuations very significantly. They also showed that the oxygen concentration was significantly lower when the magnetic field was applied.

By contrast, studies of an axial magnetic field have in general shown that oxygen levels are increased under an axial field [135] (fields of less than around 500 gauss give irreproducible results). The radial uniformity of this oxygen distribution depends on the mixing conditions in the melt and can be quite good where the mixing is optimised e.g. by the use of high rotation rates and relatively low fields. With conservative dopants however, the radial uniformity decreases progressively as the magnetic field is increased.

Initially it was believed that the higher oxygen content of crystals grown under axial fields was due to the increased crucible wall temperature necessitated by the reduced heat transport through the melt. However, subsequent work has suggested that this is not the dominant cause [136]. Rather it is a suppression of the Marangoni flow by the magnetic field which reduces the rate of evaporation of silicon monoxide which gives the higher oxygen activity in the melt. Segregation of conservative dopants appears to be well described by the segregation model of Hurle and Series [128] as shown by Ravishankar et al. [132] (Fig. 7.7).

The results obtained with a cusped field have shown good radial uniformity of oxygen distribution with a decrease in concentration with increasing magnetic field. The radial uniformity for conservative dopants such as phosphorus and carbon was also good [24, 25]. In addition there was marked reduction in melt turbulence.

The above results were obtained with the neutral plane of the magnetic field coincident with the plane of the melt surface. Subsequently Hoshikawa et al.

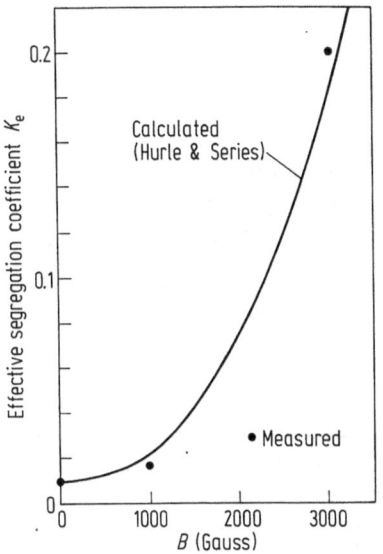

Fig. 7.7. Effective segregation coefficient of gallium in silicon as a function of axial magnetic field strength. (Ravishankar et al. [132])

[139] explored the effect of moving the neutral plane away from the plane of the crystal-melt interface. They reasoned that by doing so they would apply a vertical component of the field in the plane of the free surface of the melt and a radial component at the vertical crystal wall so that they could independently control the rate of erosion of the crucible and the rate of evaporation of silicon monoxide. By reducing the latter using a vertical component of field at the free melt surface, they were able to *increase* the oxygen concentration in the crystal. However it is to be anticipated that this would have been at the expense of radial uniformity.

7.6.2 Gallium Arsenide

Terashima and Fukuda [138] were the first to apply a magnetic field to the LEC growth of gallium arsenide. They grew two inch diameter single crystals in a transverse magnetic field of up to 1250 gauss and showed that the temperature fluctuations in the melt were reduced from a mean amplitude of around 18 °C to only about 0.1 °C. Subsequently this group developed a superconducting axial magnetic field configuration whilst a group at NTT, led by Hoshikawa [140], utilised a conventional solenoid capable of generating a field of 2000 gauss. Changes in the concentration of residual impurities and of the native defect level EL2 were observed which can only be rationalised by additionally taking account of changed crystal stoichiometry resulting from a change in the boundary layer of excess arsenic (or excess gallium) on applying the field. (see Series and Hurle [116]). As expected, radial uniformity was poor.

One unexpected benefit of the field, observed by Osaka et al. [140], was that an axial field enhanced the controllability of the crystal growth process. This was attributed to the formation of a convex growth interface in the presence of the field produced because the field gave rise to stronger radial temperature gradients in the melt. They noted that the diameter of the crystal could be controlled by varying the field. The group have extended the technique [141] to grow indium-doped crystals which are fully encapsulated in boric oxide throughout the growth. This enabled them to grow dislocation-free 5 cm diameter semi-insulating crystals.

7.6.3 Indium Phosphide

Several workers have grown either 5 cm or 7.5 cm diameter iron-doped indium phosphide single crystals in an axial magnetic field. Müller and co-workers [142] showed that, upon application of the field, the temperature in the melt changed from showing chaotic fluctuations of large amplitude to small amplitude near-periodic oscillations of period approximately 0.1 Hz when the field was applied. The irregular striations seen in crystals grown in zero field were replaced by much weaker equidistant striations corresponding to the temporal frequency of 0.1 Hz when the crystal was grown in the axial field.

The segregation coefficient of iron in indium phosphide is approximately 10^{-3} so that there is a strong macrosegregation along the length of a grown crystal. By applying a large magnetic field during the initial stages of growth and

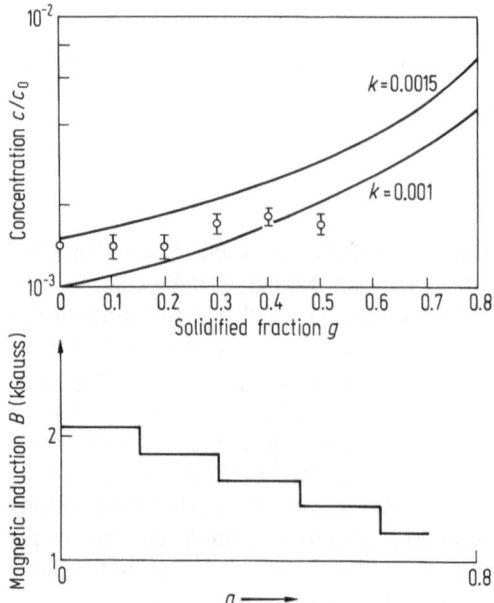

Fig. 7.8. Stepwise reduction in applied axial magnetic field performed to maintain near constancy of dopant concentration; Fe doped InP (Hofmann et al. [144]). **a** concentration profile. Curves labelled with k values are profiles expected for zero magnetic field. **b** magnetic field profile

by step-wise reducing this field as growth proceeds, one can obtain a crystal with near constant iron distribution up to a fraction solidified of around g = 0.7. This has been demonstrated by Ozawa et al. [143] and independently by Hofmann et al. [144] (see Fig. 7.8). Relevant theory has been developed by Kobayashi and Muguruma [145]. However the problem of radial non-uniformity remains.

7.7 Concluding Remarks

We have seen that the most universal effect of an applied magnetic field is the damping of convective turbulence in the melt which gives an improved micro-homogeneity of dopant distribution. In the presence of an axial field the segregation coefficient is driven strongly towards unity but at the price of poor radial uniformity. The transverse field has significant advantages for the control of oxygen incorporation in silicon but it is unlikely to find application for the III/V compounds. The advantages possessed by the cusped magnetic field make it look an attractive option both for silicon and III/V growth.

Finally a comment on other methods for damping turbulence is in order. Turbulence due to natural convective flows can be damped by reducing the Reynolds number for that flow by, for example, using a magnetic field as discussed above. An alternative approach which offers some advantage is to impose a steady forced flow which dominates over the natural convective flow. This can be achieved either by high crystal and/or crucible rotation rates or by the use of a rotating magnetic field to deliberately stir the melt. This is not in common use but was demonstrated a number of years ago for the growth of indium antimonide by Mullin and Hulme [146]. The idea has been in use for a long time in the steel industry for the electro-smelting of some grades of steel.

8 System Dynamics and Automatic Diameter Control

8.1 Introduction

In order to control the process effectively, the stability of crystal pulling against fluctuations in environmental conditions such as heater power, pull-rate etc. must be evaluated. i.e. we need to measure the dynamical response of the system to such perturbations. In the language of the electronic control engineer, the transfer function of the process must be determined. Specifically we need to know whether or not the process is stable, i.e. that, if we perturb it slightly, whether it will decay back to its original state. If the process is not stable then a servo-control system will be needed in order to maintain the desired growth conditions. In particular we will wish to grow a cylindrical crystal i.e. to ensure that the crystal radius remains constant. The response of the overall system has to be investigated but the crucial part of the process is the dynamics of the meniscus region and this is considered in the next section.

8.2 Dynamic Stability

In crystal pulling the crystal radius and its time variation are determined by the height of the supported meniscus which in turn is governed by the position of the melt isotherms. The relationship between meniscus height (h), contacting angle (Θ_L) and crystal radius (r) is given by Eq. 2.7 and is shown plotted in Fig. 8.1 for

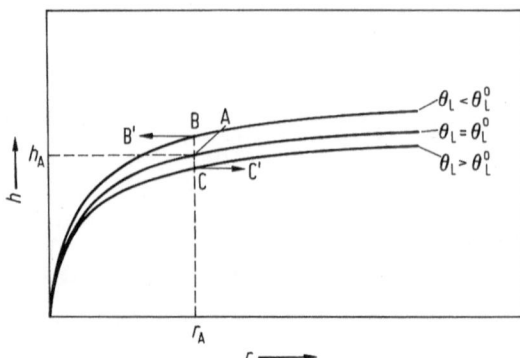

Fig. 8.1. Schematic plot of meniscus height versus crystal radius for different contact angles (Θ_L) (see text)

three different values of the contacting angle Θ_L which are respectively $\Theta_L > \Theta_L^0$, $\Theta_L = \Theta_L^0$ and $\Theta_L < \Theta_L^0$.

Suppose that, initially, the thermal field is such that the isotherm corresponding to the melting temperature is at such a height (h_A) above the surface of the bulk of the melt that the contacting angle is $\Theta_L = \Theta_L^0$ i.e. conditions corresponding to the growth of a cylindrical crystal of radius r_A. Imagine now that a perturbation in the thermal field raises the meniscus to a slightly greater height. The new contacting angle will be less than Θ_L^0 so that the crystal will begin to reduce its radius in the direction B' (see Fig. 8.1). However inspection of the figure shows that this reduction will cause Θ_L to decrease further, thus accelerating the decrease in the radius from its initial value. Similarly, a thermal perturbation which lowers the meniscus height will cause the crystal to grow with ever increasing radius in the direction C'. The pulling technique is, therefore, inherently unstable to this capillary effect as was first noted by Surek [147]. By contrast, the floating zone process is stable against such perturbations [148]. It is, in part, for this reason that servo-controllers for crystal diameter are necessary for Czochralski growth but not for floating-zone melting.

However, we have thus far considered the effects of surface tension forces alone and have taken no account of the changes in the thermal field which result from the changing meniscus shape and crystal radius. Since the meniscus height is determined by the location of the freezing point isotherm, thermal and capillary effects are strongly coupled. This coupling is investigated in the next section.

8.3 Meniscus Dynamics

The rate of growth of a crystal is coupled to the latent heat removal rate by the heat flux balance at the crystal-melt interface:

$$k_s \nabla T_s - k_L \nabla T_L = L v_n \hat{n} \qquad\qquad 8.1$$

$T_{S,L}$ is the temperature of solid and liquid respectively, $k_{S,L}$ the thermal conductivities, L is the latent heat of fusion per unit volume, v_n is the growth velocity normal to the interface and $\hat{n}$ is the unit vector normal to the interface directed outward from the crystal.

To minimise thermal stresses in the cooling crystal, experimental conditions are commonly adjusted empirically by configuring the heater geometry to give a near-planar crystal-melt interface in which case Eq. 8.1 can be written as:

$$k_s G_s - k_L G_L = L v \qquad\qquad 8.2$$

where $G_{S,L}$ are the axial temperature gradients in the crystal and melt respectively measured at the interface. G_s depends on the heat loss from the crystal which is by radiation, convection and conduction and which will, in turn, depend on ambient temperature and crystal shape. We assume that the rate of heat loss depends only on the surface area of the crystal i.e. $G_s = G_s(r)$ only. Inasmuch as the radiation view factors for the crystal are changed by changing the crystal shape and the strength of the gas convection is changed by small changes in heater power, this is only an approximation. However, it does encapsulate what is likely to be the principal effect, namely that the fraction of the heat lost radially diminishes with increasing crystal radius. The temperature gradient in the melt (G_L) will depend on the melt geometry i.e. on the meniscus height h and crystal radius r and also on the temperature at the base of the meniscus (T_B). Formally therefore:

$$v = v(r, h, T_B) \tag{8.3}$$

The growth velocity (v) is related to the pulling speed, corrected for the rate of fall of melt level in the crucible (v_p), by

$$v = v_p - \dot{h} \tag{8.4}$$

where the dot signifies time differentiation.

The angle at which the meniscus contacts the crystal Θ_L is related to the crystal radius and to the meniscus height by the Laplace-Young equation given in Chap. 2 and plotted in Fig. 8.1. Finally, the angle of taper of the crystal Θ_s is related to the time derivative of the crystal radius by the geometrical condition

$$\dot{r} = v \tan \Theta_s \tag{8.5}$$

and to Θ_L by the thermodynamic condition

$$\Theta_s = \Theta_L - \Theta_L^0 \tag{8.6}$$

where Θ_L^0 is a characteristic angle for the material in question. To establish the system dynamics we carry out a linear perturbation of these governing equations about an initial state corresponding to the growth of a right circular cylinder for which:

$$\dot{r}_0 = v_0 \tan \Theta_s^0 = 0 \tag{8.7}$$

where $v_0 = v_p - \dot{h}_0 = v_p$

is the initial growth rate, h_0 being the correct meniscus height. Small external perturbations in pulling speed and in meniscus base temperature, δv_p and δT_B are introduced and the equations linearised in these small disturbances. Thus:

$$v_p = v_p^0 + \delta v_p \tag{8.8}$$

$$T_b = T_B^0 + \delta T_B \tag{8.9}$$

These cause perturbations a and b in r and h respectively:

$$r = r_0 + a \tag{8.10}$$

$$h = h_0 + b \tag{8.11}$$

The resulting linearised equations are then:

$$v = v_0 = (\partial v/\partial r)a + (\partial v/\partial h)b + (\partial v/\partial T_B)\delta T_B \tag{8.12}$$

and
$$\Theta_L = \Theta_L^0 + (\partial\Theta_L/\partial r)a + (\partial\Theta_L/\partial h)b \qquad 8.13$$

where all the derivatives are to be evaluated at the reference state. Hence we can write, for the perturbed state, retaining only linear terms:

$$\dot{a} = v_0(\partial\Theta/\partial r)a + v_0(\partial\Theta/\partial h)b \qquad 8.14$$

Also from Eq. 8.4

$$\dot{h}_0 + \dot{b} = v_p^0 + \delta v_0 - v_0 - (\partial v/\partial r)a - (\partial v/\partial h)b - \partial v/\partial T_B)\delta T_B \qquad 8.15$$

so that

$$\dot{b} = -(\partial v/\partial r)a - (\partial v/\partial h)b + \delta v_p - (\partial v/\partial T_B)\delta T_B \qquad 8.16$$

Differentiating Eq. 8.16 and substituting it into Eq. 8.14 yields

$$\ddot{a} + [(\partial v/\partial h) - v_0(\partial\Theta/\partial r)]\dot{a} + v_0[(\partial v/\partial r)(\partial\Theta/\partial h) - (\partial v/\partial h)(\partial\Theta/\partial r)]a$$
$$= v_0(\partial\Theta/\partial h)\,\delta v_p - v_0(\partial v/\partial T_B)(\partial\Theta/\partial h)\delta T_B \qquad 8.17$$

A similar differential equation can be written for b. To proceed, expressions for the various partial derivatives in Eq. 8.17 are required.

The derivatives $(\partial\Theta/\partial r)$ and $(\partial\Theta/\partial h)$ are obtained from the solution to the Laplace-Young equation.

The thermal coefficients are formally obtained from Eqs. 8.2 and 8.3. To obtain $(\partial G_s/\partial r)$ we treat the crystal as a semi-infinite cylinder of radius r, having the plane z = 0 maintained at the melting point $T = T_M$ with the cylindrical surface radiating heat according to:

$$dT/dx + H(T - T_{amb}) = 0, \qquad 0 < z < \infty, \quad x = r \qquad 8.18$$

where x is the radial coordinate and $H = (4\varepsilon_s\sigma/k_s)T_{amb}^3$ is the linearised heat transfer coefficient, ε_s being the emissivity of the crystal. The temperature distribution is then given by Eq. 5.3.

Differentiating Eq. 5.3 with respect to the axial coordinate z yields the following approximation for the axial temperature gradient at the interface, valid for $H \ll 1$ (see Chap. 5);

$$G_s = -[dT_s/dz]_{z=0} = (2H/r)^{1/2}(T_M - T_{amb}) \qquad 8.19$$

Therefore

$$dG_s/dr = -\tfrac{1}{2}(2H/r^3)^{1/2}(T_M - T_{amb}) = -G_s/2r \qquad 8.20$$

The changes in the melt temperature gradient are the most difficult to quantify. The dominant heat transfer mechanism in the meniscus region for a molten semiconductor (having a Prandtl number much less than unity) will be conduction. If the heat loss through the crystal dominates over radiative heat loss from the meniscus, then the temperature gradient will be predominantly axial with a near planar crystal-melt interface. Thus we write:

$$G_L^0 \approx (T_B^0 - T_M)/h_0 \qquad 8.21$$

since the meniscus height will be significantly less than the thermal boundary layer thickness so that the temperature gradient can be represented linearly. In this situation:

$$\partial G_L/\partial r = 0; \quad \partial G_L/\partial h = -G_L^0/h_0; \quad \partial G_L/\partial T_B = 1/h_0 \qquad 8.22$$

Substituting these values for the derivatives into Eq. 8.17 gives the dynamics of the process. From this we can construct the transfer function of the process by taking its Laplace transform. We confine ourselves to the case where the meniscus base temperature alone is perturbed (i.e. $\delta v_p = 0$). Laplace transforming yields

$$\bar{a}/\delta \bar{T}_B = -[2k_L v_p(1 - \sin \Theta_L)]/h_0^2 L \cos \Theta_L$$
$$\times \{s^2 + [k_L G_L^0/h_0 L - \dot{v}_p h_0^2/r^2(2\beta)^{1/2} \cos \Theta_L]s$$
$$+ v_p(Lr_0 \cos \Theta_L) [k_s G_s^0(1 - \sin \Theta_L)/h_0 - h_0 k_L G_1^0/r_0(2\beta)^{1/2}]\}^{-1}$$
$$8.23$$

which is the required result.

The above is a simple model of the dynamics which relates the meniscus base temperature to the crystal radius. However, the input to the system is the power supplied to the crucible and we must further elucidate a transfer function between this input power and the meniscus base temperature.

Assuming that the crucible and crystal lose heat by radiation only, we can write an energy balance equation relating the power input P(t) to the melt temperature. For simplicity, the melt temperature is taken to be a constant, equal to the meniscus base temperature T_B. We could include transport in the melt by conduction and convection driven by melt temperature gradients but, as is shown below, these are not the dominant factors controlling the dynamics. Accordingly, the approximate energy balance equation is:

$$P(t) = \sigma[\varepsilon_h A_h + \varepsilon_L \pi(R^2 - r^2)] [T_B(t)^4 - T_{amb}^4] + (m_h C_h + m_l C_L)dT_B/dt$$
$$8.24$$

where $\varepsilon_{h,L}$ are the emissivities of heater and melt respectively, A_h is the radiating surface area of the heater, σ is Stefan's constant, T_{amb} is the ambient temperature to which the system is radiating, $m_{h,L}$ is the mass of heater and melt respectively and $C_{h,L}$ is the specific heat of heater and melt respectively.

Writing $T_B(t) = T_0 + \delta T_B(t)$ 8.25

and $P(t) = P_0 + p(t)$ 8.26

and linearising yields the disturbance equation:

$$p(t) = 4\sigma[\varepsilon_h A_h + \varepsilon_L \pi(R^2 - r^2)]T_0^3 \delta T_B + (m_h C_h + m_l C_L)\delta \dot{T}_B \qquad 8.27$$

where T_0 is the steady melt temperature obtained for a steady input power of P_0 viz:

$$P_0 = \sigma[\varepsilon_h A_h + \varepsilon_L \pi(R^2 - r^2)](T_0^4 - T_{amb}^4) \qquad 8.28$$

The transfer function is obtained by Laplace transforming Eq. 8.27 to yield:

$$\delta \bar{T}_B/\bar{p} = \{(m_h C_h + m_L C_L)[s + 4P_0/(m_h C_h + m_L C_L) T_0]\}^{-1} \quad 8.29$$

where we have neglected the term T_{amb}^4 compared to T_0^4.

Consider the propagation of this time-varying crucible wall temperature across the melt to the base of the meniscus. The above analysis is valid for temporal frequencies Ω which are small compared to $2\pi/\tau$ where τ is the characteristic time for transport of heat from crucible (or susceptor) to the base of the meniscus. For weakly convecting systems with high thermal diffusivity (for example low-melting point metals) this will be:

$$\tau = \tau_{cond} \approx R^2/\kappa \qquad 8.30$$

where κ is the thermal diffusivity of the melt ($\approx 10^{-1}$ cm^2 s^{-1}).

The time taken for heat to diffuse through the crucible wall is neglected.

For strongly convecting systems the characteristic time for transport of heat will be given by

$$\tau = \tau_{conv} \approx R/u \qquad 8.31$$

where u is a characteristic convective flow velocity in the melt. If this flow is generated by buoyancy forces then:

$$u \approx (v/R)Gr^{1/2} \qquad 8.32$$

where Gr is the Grashof number (Gr $\gg$ 1). Typically u is in the range $1-10$ cm s^{-1} so that τ_{conv} is approximately one second.

We see therefore that the dominant effect is the storage of specific heat having a transfer function given by Eq. 8.29. This fact justifies the neglect of temperature differences between heater, crucible and bulk melt.

The overall transfer function of the process is obtained by combining Eq. 8.23 and 8.29, to yield:

$$\bar{a}/\bar{p} = K/(s + p_1) (s + p_2) (s + p_3) \qquad 8.33$$

where:

$$p_1 = 4P_0/(m_h C_h + m_L C_l) T_0$$
$$p_2 = M\{1 + [1 - 4N/M^2]^{1/2}\}/2$$
$$p_3 = M\{1 - [1 - 4N/M^2]^{1/2}\}/2$$
$$M = k_L G_L^0/h_0 L - v_0 h_0^2/r^2 (2\beta)^{1/2} \cos \Theta_L$$
$$N = v_p\{k_s G_s^0 (1 - \sin \Theta_L \beta_L/h_0 - h_0 k_L G_L^0/r(2\beta)^{1/2}\}/rL \cos \Theta_L$$
$$K = -2k_L v_p (1 - \sin \Theta_L)/(m_h C_h + m_L C_l) h_0^2 L \cos \Theta_L$$
$$h_a = \{\beta^2 (1 - \sin \Theta_L)/8[1 + r^{-1}(\beta/2)^{1/2}]^3\}^{1/2}/r^2$$
$$h_\theta = -h \cos \Theta_L/2(1 - \sin \Theta_L)$$

A more complete analysis, including the growth dynamics of the conical portion of the crystal, has been given by Joyce et al. [171].

8.4 Linear Stability Analysis

The Lyapunov method of linear stability analysis can be used to examine the stability of the process whose dynamics is given by Eq. 8.33. Following Tatarchenko [149], the stability of the meniscus dynamics can be analysed by noting that crystal radius and meniscus height can be taken as independent variables. Rewriting Eqs. 8.14 and 8.16 symbolically we have:

$$\dot{a} = A_{11}a + A_{12}b \qquad\qquad 8.34$$

$$\dot{b} = A_{21}a + A_{22}b + K_T \delta T_B + K_v \delta v_p \qquad\qquad 8.35$$

The A_{ij} and K_T and K_v are coefficients which were determined in the previous section. These are given by:

$$
\begin{aligned}
A_{11} &= v_p h^2 / r^2 (2\beta)^{1/2} \cos \Theta_L \\
A_{12} &= -2v_p (1 - \sin \Theta_L) / h \cos \Theta_L \\
A_{21} &= -k_s G_s^0 / 2rL \\
A_{22} &= k_L G_L / hL \\
K_T &= k_L / hL \\
K_v &= 0
\end{aligned}
\qquad\qquad 8.36
$$

From standard Lyaponov theory the necessary and sufficient conditions for stability are:

$$A_{11} + A_{22} < 0 \qquad\qquad 8.37$$

$$A_{11}A_{22} - A_{12}A_{21} > 0 \qquad\qquad 8.38$$

These two conditions can readily be shown to be equivalent to $M > 0$ and $N > 0$. The natural frequencies of the system f_1, f_2 are given by:

$$2f_1 = -(A_{11} + A_{22}) \qquad\qquad 8.39$$

and

$$f_1^2 + f_2^2 = A_{11}A_{22} - A_{12}A_{21} \qquad\qquad 8.40$$

If f_2 is real, any perturbation decays exponentially as a damped oscillation of period $2\pi/f_2$ and decay time $1/f_1$.

If the thermal field is unaltered by the perturbation, i.e. $A_{21} = A_{22} = 0$, then stability requires

$$A_{11} < 0 \qquad\qquad 8.41$$

which is not satisfied. This is the Surek capillary instability condition described above. A necessary condition for the stability of Czochralski growth is therefore that:

$$A_{22} < 0 \qquad\qquad 8.43$$

with $|A_{22}| > A_{11}$

It has been shown [150], for the case of germanium crystal growth, that these conditions are satisfied for the analytical dynamical model described above except at extremely small crystal radii (less than 10^{-2} cm). Hence linear stability of the system is achieved by the thermal redistribution which occurs following a perturbation in melt temperature which is sufficient to overcome the instability produced by the purely capillary effect (Surek instability). This model has been experimentally tested [151] for the growth of germanium single crystals and good agreement obtained between experiment and theory. It cannot however be assumed that this result is generally applicable to all materials and all crystal pullers.

The above describes the *linear* stability of the process. It may be however that some non-linear processes are important and indeed the experimental results mentioned above prove this to be the case. It has been shown that the most important non-linear effect is the relationship between the conical angle of crystal growth and the meniscus height. A periodic modulation of the heater power produces a periodic modulation in the crystal radius but with an offset. Thus if the crystal was originally growing as a right circular cylinder, modulation of the heater power will give rise to an offset which causes the crystal to grow-out on average at some small angle whose envelope is a cone having some small positive cone angle. This has been observed experimentally [151] and modelled in numerical simulation [150] (see Fig. 8.2). The behaviour can be understood as follows.

The cone angle Θ_s which results from a modulation in the meniscus height of δh about a mean value h_0 which corresponds to the growth of a cylindrical

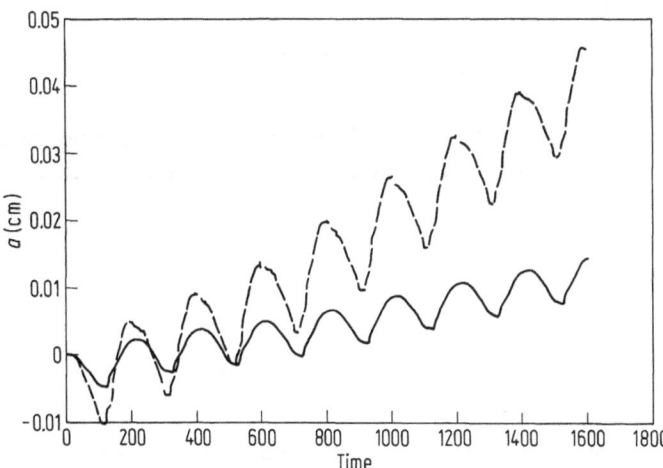

Fig. 8.2. Evolution of crystal radius for a simulated crystal, initially growing at constant radius, (a = 0), when a square wave modulation of the meniscus base temperature about its mean value in applied. Crystal radius is plotted vertically in cm. Modulation amplitude is 2 °C (*solid line*) and 4 °C (*dashed line*). [150]

crystal ($\Theta_s = 0$) is:

$$\Theta_s = A_h - B_h(h_0 \pm \delta h)^2 \qquad\qquad 8.44$$

where

$$A_h = (1 - \sin\Theta_L^0)/\cos\Theta_L^0 \qquad\qquad 8.45$$

$$B_h = [1 + (\beta/2r^2)^{1/2}]/\beta\cos\Theta_L^0 \qquad\qquad 8.46$$

We have neglected changes in r which will occur as a result of extended growth with a conical shape. From Eq. 8.15 it is seen that the natural response frequency of the meniscus to a change in height is:

$$\tilde{p} = \partial v\partial/h = k_L G_L^0/Lh_0 \qquad\qquad 8.47$$

The meniscus response can therefore be modelled by the expression:

$$h(t) = h_0 \pm \delta h\,[1 - \exp(-\tilde{p}t)] \qquad\qquad 8.48$$

If the half period of the modulation is long compared to the characteristic response time, the meniscus will have largely settled by the end of each half period of that modulation. During the half period when the meniscus height is greater than its average value, the average value of the cone angle Θ_s^+ is:

$$\Theta_s^+ = \left\{ \int_0^\tau \Theta_s\,dt - 1/v_p \int_{h_0}^{h_0+\delta h} \Theta_s dh \right\}/\tau \qquad\qquad 8.49$$

A similar expression can be written for Θ_s^- when h is less than its mean value. The integrals in Eq. 8.49 can be evaluated by substituting from Eq. 8.44 and 8.48. The overall average value of the cone angle (averaged over a whole cycle of period 2τ) will be:

$$\begin{aligned}\Theta_s &= 1/2(\Theta_s^+ + \Theta_s^-)\\ &= A_h\,\delta h^2\{h/v_p\tau - 1 + 2[1 - \exp(-\tilde{p}\tau)]/\tilde{p}\tau\\ &\quad + [1 - \exp(-2\tilde{p}\tau)]/2\tilde{p}\tau\}/h_0^2\end{aligned} \qquad\qquad 8.50$$

Inspection of this equation shows that the right hand side is greater than zero under all conditions and so, on average, the crystal progressively increases in diameter.

8.5 Automatic Diameter Control

8.5.1 Introduction

It was shown in the last section that the Czochralski process was inherently unstable to a meniscus instability but that, when the thermal redistribution which took place was taken into account then, at least for a specific set of parameters relevant to the material germanium, the system was stabilised except at the smallest crystal radii. This situation may not apply to all pullers and all materials and in any event it was also shown that an important non-linear effect gives rise to an instability. Accordingly therefore, commercial crystal pulling systems need some form of diameter control. In the following sections we briefly review the various methods for achieving this end.

It will be shown that the most successful methods are those which detect changes in the meniscus shape. The meniscus shape is characterised by the angle at which it contacts the crystal Θ_L, noting that

$$\Theta_L = \Theta_L^0 + \arctan(dr/dz) \tag{8.51}$$

The contacting angle is thus a measure of the *derivative* of the radius with respect to the growth coordinate so that, in principle, it is possible to detect a change in radius before it has occured to any significant extent. It is shown below that all of the successful servo-control methods utilise this property.

The several techniques proposed for diameter control can be classified under four headings:

1) The use of optical reflection from the meniscus.
2) Techniques for imaging the profile of the crystal.
3) Weighing the crystal or crucible.
4) The use of a die.

One method which falls outside of this classification was proposed by Vojdani et al. [152] who reported on the use of the Peltier effect to control, by open loop, the diameter of pulled germanium crystals. The advantage of the Peltier effect is of course that heat can be generated or removed exactly at the growth interface so that the control of the growth rate can be achieved without thermal lag. However the method is limited to those very few materials which have an adequate electrical conductivity to avoid excessive Joule heating and have a large Peltier coefficient – i.e. it is limited to the small and medium band-gap semiconductors.

8.5.2 Reflection from the Meniscus

The first reported operational automatic diameter control system, for the growth of silicon crystals from 1 kg melts, was by Patzner et al. [153]. These authors used an optical pyrometer operating in the spectral region 0.65–0.9 µm focused on the melt surface adjacent to the crystal. The pyrometer output was used to modulate crystal pull rate and/or crucible lift rate. Similar techniques were subsequently described using computer control and modulation of heater power as well as pulling speed. The photodiode was kept in focus on a "bright ring" on the meniscus by translating the crucible at a rate to compensate for the fall in melt level (Fig. 8.3). Initially attributed to latent heat evolution, this bright ring is in fact a reflection, in the meniscus, of the hot regions at the top of the crucible. A change in meniscus shape therefore displaces this image so that the

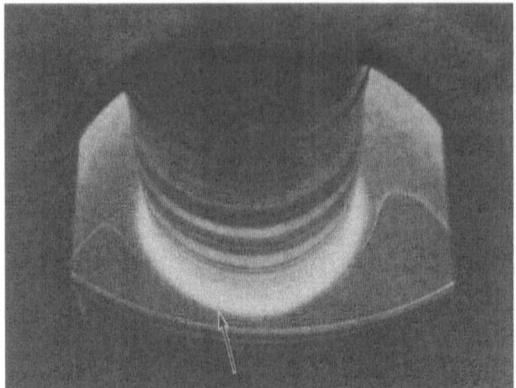

Fig. 8.3. Silicon crystal growth showing the 'bright ring' (*arrowed*). (Digges et al. [22])

system is in fact detecting insipient changes in crystal radius by detecting changes in the meniscus slope. It is a highly successful and widely used technique for silicon, to which it is well suited because of the high Laplace constant of molten silicon (resulting in a large meniscus height) and its high optical reflectivity. An alternative "active" meniscus reflection system has been described by Gross and Kersten [154] in which a laser beam is reflected vertically at near-normal incidence from the melt surface. The choice of near-normal incidence obviates the problem of the falling melt level but reduces the sensitivity of the technique since the point of observation is remote from the three-phase boundary. The technique was successfully used to grow crystals of potassium chloride and bismuth germanate but suffered from problems of noise caused principally by gravity waves on the melt surface.

8.5.3 Crystal Imaging

Such methods are based on the direct measurement of crystal diameter from an imaged profile of the crystal. Imaging systems working at optical, infra-red and X-ray frequencies have all been developed. T.V. imaging systems using a conventional vidicon have been described by Gartner et al. [155] and infra-red imaging by O'Kane et al. [156]. Both techniques suffered from the fact that the crystal could not be viewed normal to its axis and required the use of relatively large diameter crucibles in order to be able to image the interface in the later stages of growth when the melt level was low.

Since the crystals could not be viewed at normal incidence, the crystal-melt interface appeared as a half ellipse. Selection of the scan line on the T.V. image corresponding to the interface position in order to measure the current crystal diameter was therefore difficult. However the method does have the advantage that control is possible from the point of seed-on onwards so that the critical shoulder region of the crystal can be controlled automatically.

Uelhoff [260] has developed a sophisticated optical imaging technique for the computer controlled growth of metal and metal alloy single crystals. This involves imaging the three-phase boundary using a mirror system to minimise fogging of the optical components.

Two X-ray imaging systems have been reported; both were applied to the growth of gallium phosphide by the liquid encapsulation technique. Pruett and Lien [157] used only manual control assisted by an X-ray image obtained by using a small X-ray source and image intensifier. They mounted beryllium windows in the pressure chamber walls replacing the copper RF coil with an aluminium one because of the high X-ray absorption of the former. Van Dijk et al. [158] employed a similar construction but with aluminium windows instead of beryllium ones (Fig. 8.4). These X-ray systems have the advantage of viewing the crystal at normal incidence and the scan line corresponding to the plane of the interface can be selected automatically by counting a predetermined number of lines above the line corresponding to the melt level remote from the crystal. In practice it was found necessary to select a line corresponding to a plane below the crystal-melt interface.

The basic weakness of all of these imaging techniques is that they measure the change in diameter only after it has occurred by a significant amount and for a significant time. The problem is particularly acute with optical and IR imaging where the crystal cannot be viewed at normal incidence. For the reasons given above, it is surely better to monitor the width of the meniscus at some fixed distance below the crystal-melt interface since this provides some measure of changes in the contacting angle. This would appear to have been appreciated by Van Dijk et al. who reported that their best results were obtained when they selected a line corresponding to the upper portion of the meniscus.

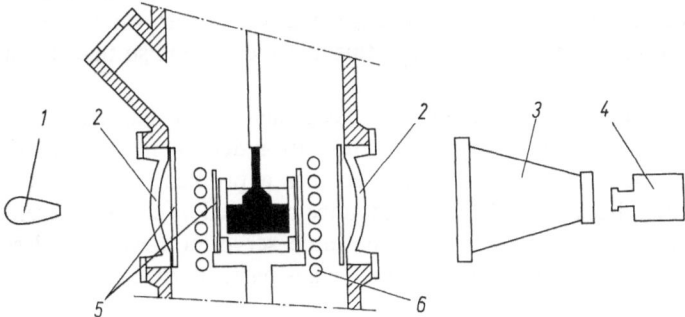

Fig. 8.4. Cross-sectional view of an X-ray imaging system fitted to a high pressure LEC puller for the growth of gallium phosphide. (*1*) X-ray source, (*2*) aluminium windows, (*3*) X-ray image intensifier, (*4*) TV camera, (*5*) radiation shields and (*6*) RF aluminium coil. (van Dijk et al. [158])

8.5.4 Crystal and crucible weighing

The use of a servo-controlled crystal weighing system was first reported by Bardsley et al. [159]. They compared an observed weight with a demanded weight and used the error signal to modulate the heater power. They demonstrated an ability to grow long single crystals of sodium chloride (a low density material) with closely controlled diameter. They discovered however, that the system was unstable during the growth of germanium crystals at slow growth rates (less than several centimetres per hour) and they identified the cause of this instability which is outlined below. This work was followed by a number of publications from many laboratories utilising either crystal and crucible weighing, almost exclusively for the growth of oxide single crystals (For a review see Hurle [160]).

The problem of the instability observed by Bardsley et al. delayed application of the technique to semiconductors until this problem was overcome in subsequent work at R.S.R.E., Malvern. The group developed a crucible weighing assembly for the use in the pressure chamber of a liquid encapsulation pressure puller and have successfully demonstrated the automatic growth of indium phosphide, gallium phosphide and gallium arsenide single crystals (Bardsley et al. [172]). The choice between crystal weighing and crucible weighing is a fine one. Crystal weighing in a pressure puller requires that the load cell be protected from the strong rising convective currents and corrosive evaporants. Crucible weighing, where RF heating is employed, carries with it the problem that a change in the induced RF current in general induces a vertical force on the crucible for which compensation has to be provided.

Two modes of control are possible with the weighing technique: either the actual weight is compared with a demanded weight and the difference utilised as an error signal or the weight is differentiated and compared with the demanded rate of increase of weight, the difference forming the error signal. The former ("weight" mode) seeks to keep the total weight of the crystal at its demanded value and so compensates for any error in diameter at a previous stage of growth by causing an error of opposite sign. Thus "start-up" errors can produce a damped oscillation in shape which is propagated down the growing crystal. However, the signal-to-noise performance of this method is good so that it can be used down to very low growth rates.

The "differentiated weight" mode seeks to keep the current diameter at its correct value, ignoring past history. Signal differentiation does increase the noise level of the signal markedly but this only becomes a problem at very slow growth rates. Further, because of the phase advance provided by signal differentiation, this mode tends to provide a more stable servo-loop which is less sensitive to the thermal lags in the equipment. This is particularly significant at high growth rates.

However, the relationship between rate of change of crystal weight and radius is not a simple one because the load cell measures not only the static weight of the crystal but also the surface tension forces which act on it. This renders the process dynamics considerably more complicated and this is reviewed in Sect. 8.6 below. It is in fact the cause of the instability first seen in germanium crystal growth at low speeds mentioned above.

The problem has been overcome by Bardsley et al. [161], [162], as described below, which has enabled the technique to be applied to the commercially important semiconductors germanium, gallium arsenide, gallium phosphide and indium phosphide. The presence of a boric oxide encapsulant during growth of the III/V compounds does not present a problem. (Effects due to viscous drag and surface tension of the encapsulant appear to be of secondary importance only).

Through the surface tension effects at the 3-phase boundary, the force experienced by the load cell is a function of the meniscus angle Θ_L and so the error signal contains a term which is a measure of this and hence of the derivative of the radius error. This term is in effect sensing change in meniscus shape and thereby anticipating impending diameter change.

8.5.5 The Use of Dies

The use of a die floating on the melt surface or rigidly attached to the crucible has been widely researched but principally for the growth of ribbon crystals. This is outside the scope of this monograph. Two classes of die have been employed – wetting and non-wetting. Here a brief review of the use of both types of die for the growth of cylindrical crystals is given.

The use of non-wetted dies has been extensively explored in Russia following pioneering work by Stepanov [163]. In work reported in 1959, but performed much earlier, he described the fabrication of metal strips, tubes and bars by pulling through a die floating on, or fixed in the vicinity of the surface of a bath of molten metal. Subsequently he considered the application of the process to the growth of single crystals. Antonov [164] reported specifically on the growth of cylindrical crystals of germanium and indium antimonide using this technique and the geometrical arrangement of the die and crucible is shown in Fig. 8.5. An excess pressure is generated to force the liquid to the top of the die

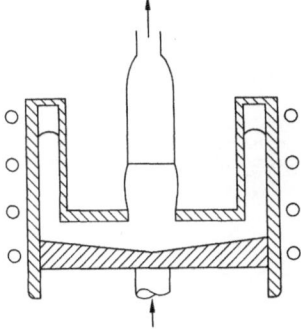

Fig. 8.5. Crystal pulling through a die. [164]

and to control the meniscus shape. It has been shown that there is an optimal pressure for which the crystal diameter is sensibly independent of small fluctuations in meniscus height: i.e. it is stable by the Surek criterion. Development of the technique for the growth of germanium has been reported by Dudnik et al. [165], Smirnov [166] and Egorov et al. [167]. The dynamic stability of the Stepanov process has been comprehensively analysed by Tatarchencko [168].

The use of wetted dies was pioneered by LaBelle and Mlavsky [169] with their edge-defined, film-fed growth (EFG) technique. This has been used to produce sapphire crystals of a variety of shapes (Fig. 8.6) and was subsequently developed for the production of silicon ribbons. Although applicable to the growth of cylindrical crystals, it does not appear to have been used to any extent for this purpose.

In Stepanov's method, the lower end of the meniscus is anchored at the edge of the die. Diameter control can also be effected by confining the lower end of the meniscus within a slightly conical surface which is unwetted. Indeed, the wall of the crucible can act in this way to provide a self-stabilisation of the diameter of crystals whose diameter has become comparable with that of the crucible. The use of a die with a conical contacting surface applied to the growth of gallium arsenide and gallium phosphide by the liquid encapsulation technique has been reported by Cole et al. [170]. The die was made of silicon nitride and termed a "coracle". It was floated on the melt beneath a layer of boric oxide encapsulant. Wetting of the silicon nitride by the boric oxide ensures non-wetting (zero contact angle) by the molten gallium phosphide or gallium arsenide. It can be shown that under these conditions growth takes place in a stable regime where very precise control of temperature is unnecessary. The technique works well for crystals grown on a [111] axis but, for reasons not well understood, gives a very

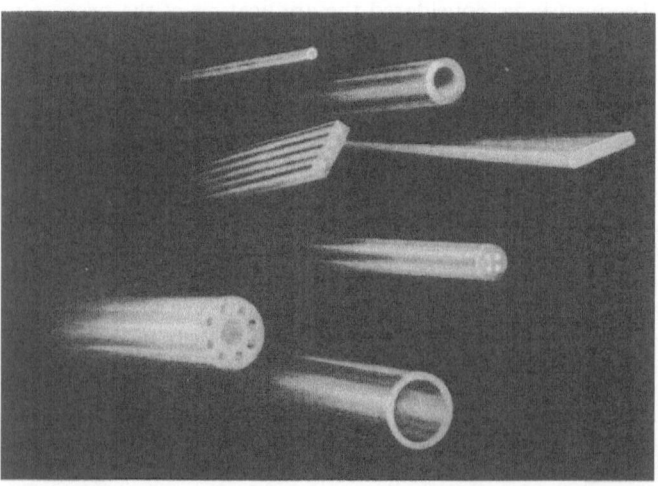

Fig. 8.6. Single crystal sapphire forms grown using dies. (LaBelle and Mlavsky [169])

high incidence of twinning when growth is attempted on a [100] axis. A further limitation of the method is that manual control is necessary for the initial stages of growth until the final diameter has been obtained.

The powerful advantage of both wetting and non-wetting dies is that, used under the right conditions, they avoid the Surek-type instability which is a feature of conventional Czochralski growth.

8.6 Dynamics of Control by the Weighing Technique

The weighing technique is the most widely applicable method of automatic diameter control i.e. usable over the widest range of materials. As already explained, imaging techniques are limited by the inability to look into the crucible at normal incidence. Optical techniques do not work well with melts of low reflectivity. The use of dies requires that materials, which have chemical compatibilty with the melt and the required non-wetting properties, have to be found. Weighing is carried out remotely from the melt and requires only some mechanical modification to the pulling system. However, the relationship between weight and radius is complicated and, as is shown below, is in some cases ill-conditioned.

The derivation of a functional relationship between weight and radius follows. We consider specifically the weighing of the crystal but, with appropriate change, the analysis carries through equally for the case of crucible weighing. The force experienced by a load cell on which the pull rod and crystal are suspended is comprised of the static weight of that pull-rod and crystal plus contributions due to the vertically resolved component of surface tension exerted along the length of the three-phase boundary and also to the "weight" of the supported meniscus. The measured force is thus:

$$F(t) = m_0 g + \int_0^t \rho_s g \pi r^2 v \, dt + \pi r^2 \rho_L g h + 2\pi r (\gamma_0 - \gamma_\theta \Theta_s) \qquad 8.52$$

where m_0 is the mass of pull rod plus crystal at time $t = 0$ and where γ_0 and γ_θ are given by:

$$\gamma_0 = \sigma_{LG} \cos \Theta_L^0$$

and

$$\gamma_\theta = \sigma_{LG} \sin \Theta_L^0$$

If we seek to grow a cylindrical crystal of radius r, then the measured force should increase linearly with time according to:

$$F_{ref} = m_0 g + \rho_s g \pi r_0^2 v_0 t + \pi r_0^2 \rho_L g h_0 + 2\pi r_0 \gamma_0 \qquad 8.53$$

where h_0 is the meniscus height corresponding to a radius r and contacting angle Θ_L^0

Differentiating this weight signal, we see that:

$$\dot{F}(t) = \rho_s g \pi r^2 v + \pi \rho_L g r (2\dot{r}h + r\dot{h}) + 2\pi[\dot{r}\gamma_0 + \dot{\gamma}_0 r] \qquad 8.54$$

If the radius deviates from r by a small amount, thus changing the contact angle by a small amount Θ_s, then we can linearise the force equation in these small deviations to obtain an expression for the deviation in the measured differential of the force from its expected value. This is:

$$\dot{H} = a + \eta \dot{a} - \lambda \ddot{a} \qquad 8.55$$

where η and λ are given by Eq. 8.59 below.

$$\dot{H} = (\dot{F} - \dot{F}_{ref})/2\pi g r_0 v_0 \rho_s \qquad 8.56$$

is a measure of the error in the differential of the weight signal. This is used to provide a control signal.

Equation 8.55 is the dynamical equation describing the growth of the cylindrical portion of the crystal. The equivalent expression for the growth of the conical region is somewhat more complex [171]. The transfer function of this part of the overall process is obtained, for cylindrical growth, by taking the Laplace transform of Eq. 8.55 which yields:

$$\bar{H}/\bar{a} = 1 + \eta s - \lambda s^2 \qquad 8.57$$

where s is the Laplace operator.

Combining this result with that obtained in Sect. 8.3 we obtain, for the transfer function of the overall process:

$$\bar{H}/\bar{p} = K'(s + z_1)(s + z_2)/(s + p_1)(s + p_2)(s + p_3) \qquad 8.58$$

where:

$$z_1 = \eta\{1 + [1 + 4\lambda/\eta^2]^{1/2}\}/2\lambda$$

$$z_2 = \eta\{1 - [1 + 4\lambda/\eta^2]^{1/2}\}/2\lambda$$

$$\eta = [2(\rho_L h_o + \gamma_o/rg) - (\rho_s - \rho_L) h_a]/2\rho_s v_0 \qquad 8.59$$

$$\lambda = [2\gamma_\theta/rg - (\rho_s - \rho_L) h_\theta]r/2\rho_s v_0^2$$

The coefficients η and λ have been obtained from an approximation (Hurle [18]) to the equation for the meniscus profile (Eq. 2.7).

Inspection of the parameters shows that $z_1 < 0$ when η/λ is positive. This renders the system inherently unstable. An alternative way of looking at the problem is to note that the complementary function of the dynamical equation (Eq. 8.57) is a growing exponential with exponent $\eta/2\lambda > 0$.

Physically this effect comes about as follows. Consider a crystal growing initially as a right circular cylinder supporting a meniscus of height h_0 with that meniscus contacting the crystal at an angle Θ_L^0 to the vertical. If a fluctuation in temperature raises the height of the meniscus slightly, then the contacting angle decreases and, as growth proceeds, the crystal radius will decrease. The ultimate effect of this will be a decrease in the rate of increase of the crystal weight but the immediate effect is to cause the surface tension vector to swing toward the vertical so that the vertically-resolved component of that force (which is measured by the load cell) will increase. Furthermore, the increase in meniscus height can be thought of as replacing a wafer of crystal of height δh by an equivalent wafer of melt. Since, in the case of the zinc-blende and diamond-cubic semiconductors, the melt has a higher density than the crystal then the apparent weight of the crystal has increased on this account also.

Therefore the first effect of a heater power increase, which will ultimately lead to a crystal of reduced radius, is to give rise to a change in the rate of increase of weight of the "wrong" sign so that any simple proportional controller will change the heater power in the wrong direction resulting in the servo performing a limit cycle oscillation. This is illustrated schematically in Fig. 8.7 which shows the change in radius, meniscus height and force as a function of time following a small step increase in heater power.

This "anomalous" dependence of weight differential on radius was overcome by Bardsley et al. [162] in the following manner. The servo system which they developed is shown in Fig. 8.8. The "anomalous" surface tension contribution to the weight term is estimated using an estimate of the current radius of the crystal and this is subtracted from the measured force so that the resultant signal is well behaved. This procedure is equivalent to removing an estimate of the term $(s + z_1)$ from the numerator of Eq. 8.58. With reference to Fig. 8.8, this is achieved by using the estimated radius error, taken at point 16, differentiating it (13) to provide an estimate of a, which is fed through a variable attenuator which provides a scaling for the quantity η in Eq. 8.57. The resulting signal is then

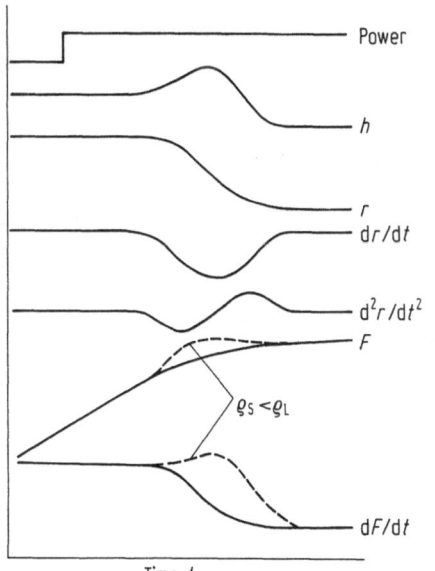

Fig. 8.7. Change of meniscus height, crystal radius and its derivatives and force as a function of time following a step change in heater power

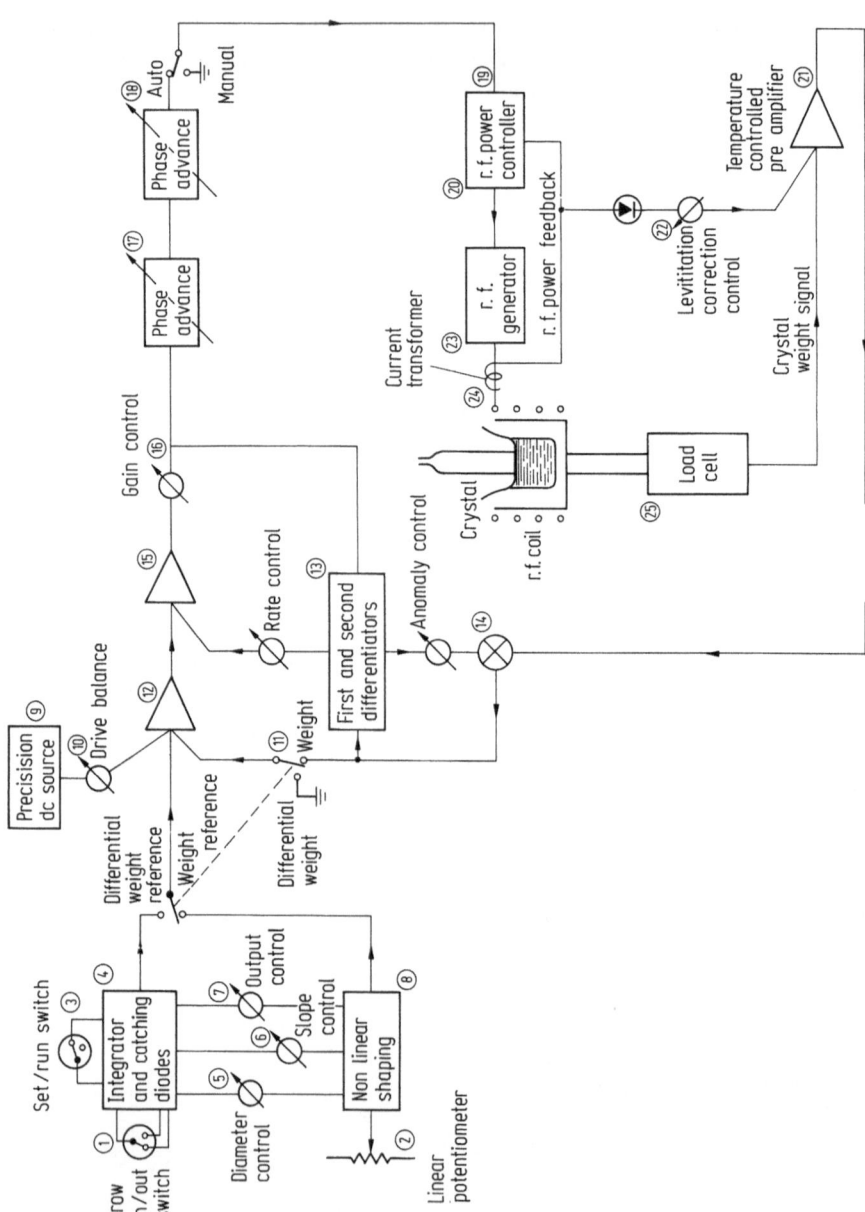

Fig. 8.8. RSRE-designed servo-control system based on measurement of crucible weight. [162]

added to the weight signal at 14. It is further differentiated at 13 before comparison with the reference signal at 15. The effect of this strategy on the control is dramatic. Without this "anomaly control", the servo performs violent limit cycle oscillations at all but the fastest growth speeds; with it, well-controlled single crystals can be obtained, see Fig. 8.9.

One further feature of the servo-controller shown in Fig. 8.8 requires mention. The "levitation corrector control" is provided to correct for spurious changes in the apparent weight of the crucible which are caused by the vertical thrust generated between induced and inducing RF currents when induction heating of the crucible is employed. Correction is effected by rectifying a sample of the RF power and adding it to or subtracting it from the weight signal as appropriate. The correction is made via a variable attenuator which is set so that no change in corrected weight is seen when an incremental change in power about the mean level is made. This controller is utilised in the Cambridge Instrument Ltd (now Metals Research Semiconductors) range of pressure pulling systems.

A limitation of the weighing method applied to the LEC growth of III/V compound semiconductors is evident in the grow-out phase from seed to final diameter. As the crystal grows out, it does so with a convex crystal-melt interface such that, like an iceberg, it is essentially floating on the melt and grows without

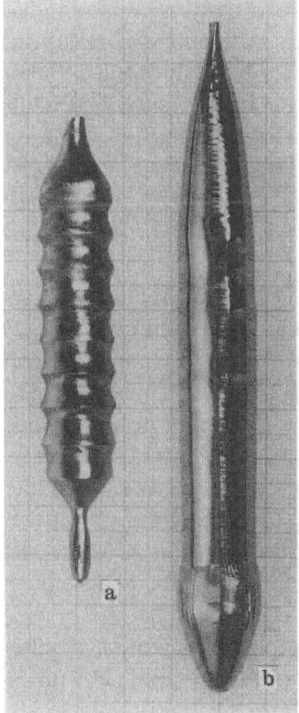

Fig. 8.9. Germanium single crystals grown under automatic control (*a*) with simple PID control only and (*b*) with "anomaly control" (right hand crystal)

apparent increase in weight. By the time that a change in weight signal is detected, the process of grow-out may have accelerated to the point (called flash-out) where it can no longer be controlled. This requires manual intervention and can result in a crystal whose grow-out profile is a set of sequential cusps. Such a profile gives rise to high concentrations of elastic stress in the crystal as it cools (see Chap. 10) and is inimical to the attainment of a low dislocation density in the crystal. To overcome this problem one must either dispense with weight as a sensor signal or one must find some more sophisticated way in which to use this sensor in order to provide a robust control signal during the grow-out phase. Both approaches have been attempted for the control III/V semiconductor growth as follows:

The bright ring technique is being developed [173] using sophisticated averaging techniques to cope with the distortion of the image of the bright ring when viewed through a boric oxide layer which contains gas bubbles which are carried around by the rotating crystal. The present author and colleagues [171] have followed the other route and devised a technique whereby the anomalous surface tension term in the weighing method, instead of being an unwanted signal which one attempts to remove, is utilised to give information about the cone angle. This is achieved by modulating the heater power with (say) a square wave modulation. The amplitude is chosen to modulate the meniscus height by some few percent which in turn modulates the contacting angle Θ_L through some few degrees. By cross-correlating the output of the weight differentiator with the square wave input drive, this modulation of the contacting angle can be detected. The functional dependence of the magnitude of the cross-correlation on the system parameters can be obtained algebraically. This results in an extremely complex expression containing the growth parameters such as crystal radius, cone angle, growth speed, temperature gradient etc. [171]. Taking material parameter typical of a semiconductor and plotting this relationship, it has been shown that the maximum amplitude of the cross-correlation signal (A_{max} which occurs for a delay of τ_{max} is a well behaved function of radius and cone angle.

The time evolution of A_{max} during the growth of a crystal shoulder of any predetermined shape can be calculated and stored in a look-up table and used to generate the reference control signal for a servo-controller. This new control strategy is capable of giving better control of the grow-out phase. A disadvantage is of course that by deliberately modulating the heater power, one is modulating the growth rate and hence inducing striations into the crystal. However, by changing over progressively to the weight differential servo as the shoulder of the crystal is rounded and simultaneously decreasing the amplitude of the square wave modulation to zero, one limits the striations to the cone region. Since, in commercial operation, the cone is rejected, no disbenefit is incurred.

It can be seen that the servo-control of semiconductor crystal pulling is quite a sophisticated task and one which has only become fully possible with a thorough understanding of the process dynamics.

Design of a suitable servo-controller is a specialist branch of electrical engineering (see for example Dorf [225]). However, some basic principles relevant to crystal growth can be given. Confining considerations to the weighing technique, control can be effected by weighing either the crystal or the crucible. With crystal weighing the problems which arise are a) protecting the load cell from rising convected heat and corrosive evaporants and b) weighing a rotating object (the pull rod). The former problem is exacerbated in a pressure puller by the fact that the Grashof number, which determines the degree of ambient gas turbulence, increases roughly with the square of the system pressure. With crucible weighing, the load cell can be protected in a separate chamber beneath the growth chamber but the static weight to be tared off (the weight of the crucible, its contents and the support assembly) can be large relative to the final crystal weight, making it more difficult to obtain the required weight sensitivity. In systems where the crucible is rotated there is again the additional problem of weighing a rotating object.

Very long thermal lags can be present with large resistance heated pullers which can result in instability in the servo-control. This is overcome by incorporating a phase advance network into the controller. This involves differentiating the error signal.

To obtain high accuracy in the controlled variable (the crystal radius) under steady state conditions (i.e. to avoid drift away from the set point which would result in a slightly tapered crystal), it is necessary to work with a high servo-loop gain. However, this strategy can lead to instability, particularly in a noisy system such as a crystal puller. The remedy is to provide compensation via a feed-forward network (which is essentially an integrator). This confers the additional benefit of attenuating the high frequency response of the system, and hence reducing its sensitivity to noise.

At this point one must ask what is the objective of the use of servo-control. The attainable tolerance on crystal radius can be improved by tightening the servo-response but this will, in general, be at the price of rapid fluctuations in the microscopic growth rate as the power is cycled up and down to maintain the crystal radius accurately at its demanded value. This will result in a crystal which has pronounced dopant striations which may well be deleterious for the crystal application. Various optimal control strategies can be devised. One possible one would be to minimise the fluctuations in crystal radius, but subject to not exceeding some maximum excursion in the microscopic growth velocity, or alternatively, to avoid conditions which lead to melt-back during any part of the oscillation cycle since this has been shown (Chap. 6) to be particularly deleterious in respect of striation generation.

A more sophisticated approach utilises more than one control variable. Satunkin and Leonov [232] have considered the growth of non-anomalous materials (i.e. nonsemiconductors) with control of both heater power and pulling speed. Employing an optimisation criterion which seeks to minimise the mean square error in the crystal radius, they deduced that the best performance was obtained by using standard PID (*P*roportional-*I*ntegral-*D*ifferential) control of the heater power with PD control of the pulling speed. Control of the latter provides a fast-acting response with control of heater power being used to maintain mean power at a value appropriate to the demanded radius. This strategy has also been employed for silicon crystal growth by Kim et al. [233].

Anomalous materials bring additional problems, as has been indicated above, and the practical solution of these has been described by Bardsley et al. [162]. The presence of a liquid encapsulant also complicates the use of a weight sensor since it exerts a bouyant force on the crystal the magnitude of which depends on the volume of crystal which is immersed which in turn depends on the crystal size and shape. It is necessary to correct for this when computing the rate of change of weight required to give the desired radius. Riedling [234] has reported on the development of a control system having this capability.

Gevelber et al. [235] have provided design criteria for advanced control systems which seek to optimise control, not simply of crystal radius, but also of dopant uniformity and defect concentration. However, effective implementation of these criteria is likely to require a much fuller understanding of the dynamics of the overall process including aspects of segregation and of stress generation in the cooling crystal.

Thermal conditions can change markedly as growth proceeds and the melt level in the crucible falls. Thus what may have been an optimum set of control parameters (gain etc.) during the initial stages of growth may be far from optimum toward the end of the growth run. This indicates that adaptive control is desirable in which the system continually monitors the response function and adapts the control algorithm to maintain optimal control. This requires digital implementation with a non-trivial amount of computing power.

9 Morphological Stability of a Planar Rotating Interface

9.1 Introduction

Inhomogeneity of chemical composition on a macroscopic scale resulting from the segregation of solute at the crystal-melt interface during solidification of a finite melt has already been considered. Additionally it has been shown that a macroscopic radial non-uniformity can result from flow conditions under which the solute boundary layer does not remain embedded within a momentum boundary layer flow characterised by the rotating crystal. Such a situation can occur for example in the presence of an axial magnetic field. Further, we saw in Chap. 6 how the response of the interface to oscillations of zero wave number, (i.e. plane wave oscillations) gives rise to striations.

There is however a further form of instability which can be a major cause of quite massive inhomogeneity in alloy single crystals and compound crystals grown from melts having a non-congruent composition. This occurs when the macroscopically-planar or smoothly curved interface becomes unstable to a perturbation in shape on the scale of the lateral diffusion length for solute in the melt (D/v_p where D is the Fickian diffusion coefficient). This occurs under conditions referred to as constitutional supercooling which is now explained. This is a universal phenomena in melt growth and perhaps the single most important stability phenomenon affecting the perfection of melt-grown crystals. Attention in this book is confined to a desciption of the concept of constitutional supercooling and the specific effects of crystal rotation on its occurrence.

9.2 Constitutional Supercooling

Supercooling resulting from the redistribution of solute at the crystal-melt interface (so-called constitutional supercooling) was pointed out by Rutter and Chalmers [175] and the theory was elucidated by Tiller et al. [176] for the case of an unstirred melt. The validity of this theory has been extensively tested experimentally. As already discussed, when growth proceeds into a doped melt containing a solute whose segregation coefficient is less than unity, solute is continually being rejected into the melt at the interface resulting in a layer containing an enhanced solute concentration immediately adjacent to the interface. Solute in this layer is removed into the bulk of the melt by diffusion close to the interface and by fluid flow at greater distances. (For a solute with segregation coefficient greater than unity, there is a *depletion* layer in the melt ahead of the interface). The thermodynamic freezing point of the alloy melt varies through the boundary layer, being lowest at the crystal-melt interface. This is shown in Fig. 9.1 for a solute with segregation coefficient $k < 1$ which corresponds to a system having a phase diagram such as that shown schematically in Fig. 6.1a.

The liquidus curve on this phase diagram is downward sloping so that increasing solute concentration represents decreasing freezing temperature. The addition of salt to water is a familiar example. The freezing point distribution corresponding to the solute distribution in the boundary layer therefore has the form shown in Fig. 9.1. The actual temperature at the crystal melt boundary will be at, or slightly below, the equilibrium temperature, the amount of undercooling beneath the thermodynamic equilibrium temperature being equal to that required to drive the kinetic processes of crystallisation at the imposed rate. In the absence of facets on the crystal-melt interface, this undercooling is in general negligible at the usual growth rates. Consider now two possible actual

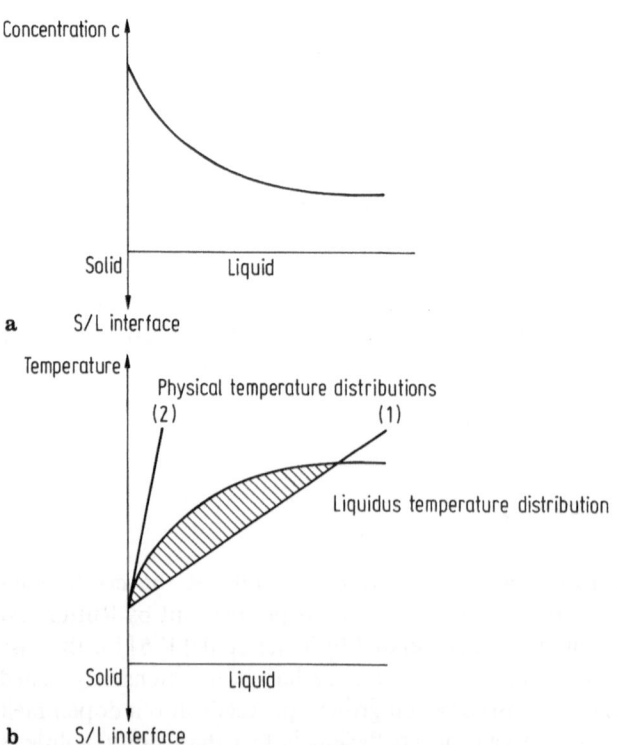

Fig. 9.1. a Solute concentration distribution ahead of a crystallising interface for a solute with $k < 1$.
b Resulting liquidus temperature distribution. A zone of constitutional supercooling (*shaded*) exists if the actual temperature gradient is sufficiently small (line (*1*))

temperature distributions in the melt designated by the lines 1 and 2 in Fig. 9.1b. Both correspond to a positive temperature gradient but, with the steeper gradient, the actual temperature in the melt is everywhere greater than the liquidus temperature except actually at the crystal-melt interface. However, with the shallower gradient, there exists a region (shaded) for which the actual temperature is below the liquidus temperature, i.e. that region of melt is supercooled. The name constitutional supercooling is given to this phenomenon. The criterion for the existence of a zone of constitutional supercooling (S) in the melt is:

$$dS/dz = m \, dC_L/dz - dT_L/dz > 0 \qquad \text{at } z = 0 \qquad\qquad 9.1$$

where m is the slope of the liquidus curve on the phase diagram. The interface is deemed to lie in the plane z = 0.

Constitutional supercooling is undesirable because it produces instability in a planar crystal-melt interface. A naive explanation of this is as follows:

If, by a chance fluctuation, a protuberance forms on the crystal itself, it finds itself in a region of supercooling and hence can grow more rapidly. In doing so, it will segregate solute laterally in the plane of the interface, making the liquid adjacent to neighbouring regions of interface more solute-rich and hence with a lower freezing point. The solidification of these regions therefore tends to be inhibited and the protuberance is thereby stabilised. A number of such projections can be expected to form in a close packed array separated by distances related to the lateral diffusion distance (D/v_p) – see Fig. 9.2. The regions between these projections will become progressively richer in solute and long solute-rich channels will form in the crystal.

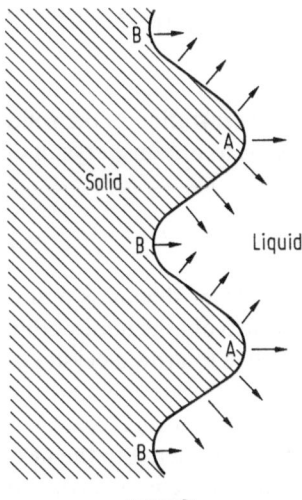

Macroscopic growth direction

Fig. 9.2. Schematic representation of the formation of a cellular interface. *Arrows* signify rejection of solute by the crystallising interface; in consequence the solute concentration rises ahead of regions B relative to regions A; this stabilises the cellular structure

This broadly is what is observed experimentally and is referred to as a cellular structure. We next derive conditions under which the zone of constitutional supercooling can exist in a Czochralski melt. In Sect. 9.3, examples of cellular structure in oxide and semiconductor crystals and the concommittant chemical and structural inhomogeneity are described. In Sect. 9.4, the rather hand-waving concept of instability as being due to the existence of a state of constitutional supercooling is replaced by a proper stability theory.

To evaluate Eq. 9.1 we must find expressions for the quantities dC/dx and dT_L/dx. The latter is simply obtained by writing the heat flux balance equation at the crystal-melt interface:

$$k_L dT_L/dz = k_s dT_s/dz - Lv_p \qquad 9.2$$

Taking the temperature gradient in the solid dT_s/dz at the interface as a fixed quantity $(= G_s)$, determined by the ambient cooling conditions (see Chap. 5), we see that the temperature gradient in the melt varies as the growth speed is varied due to the contribution of latent heat to the overall heat balance. The concentration gradient at the interface is obtained from the Burton, Prim and Slichter [97] analysis of segregation at a rotating disc. The following flux balance equation can be written at the interface:

$$D \, dC_L/dz - v_p[C_s - C_L(0)] = 0 \qquad 9.3$$

Using the Burton, Prim and Slichter equation (Eq. 6.10) to relate the concentrations in solid and liquid at the interface to the far-field liquid solute concentration, we obtain:

$$D \, dC_L/dz = - v_p C_L(1 - k) \{k + (1 - k) \exp(- v_p \delta/D)\}^{-1} \qquad 9.4$$

Inserting this into Eq. 9.1 gives the condition for the existence of a zone of constitutional supercooling:

$$[dS/dZ]_{z=0} = - mv_p(1 - k) \{[k + (1 - k) \exp(- v_p \delta/D)]\}^{-1}$$
$$- (k_s G_s - Lv_p)/k_L > 0 \qquad 9.5$$

Alternatively, this can be expressed in terms of the concentration in the solid – which is a more appropriate parameter if the objective is to obtain a crystal of a given solute concentration. The result is:

$$[dS/dZ]_{z=0} = - mv_p C_s(1 - k)/Dk - (k_s G_s - Lv_p)/k_L \qquad 9.6$$

which is seen to be independent of the solute boundary layer thickness δ and hence of Ω, the crystal rotation rate. Thus, one sees that to avoid the undesirable condition of constitutional supercooling, given the objective to grow a crystal of a given solidus concentration C_s, one must keep the growth rate low and the temperature gradient in the melt high. This latter condition exacerbates the problem of thermal stress in the cooling crystal (which is considered in the next chapter). Note that solutes with small segregation coefficient are most prone to give rise to a condition of constitutional supercooling. Melt-stirring has no effect

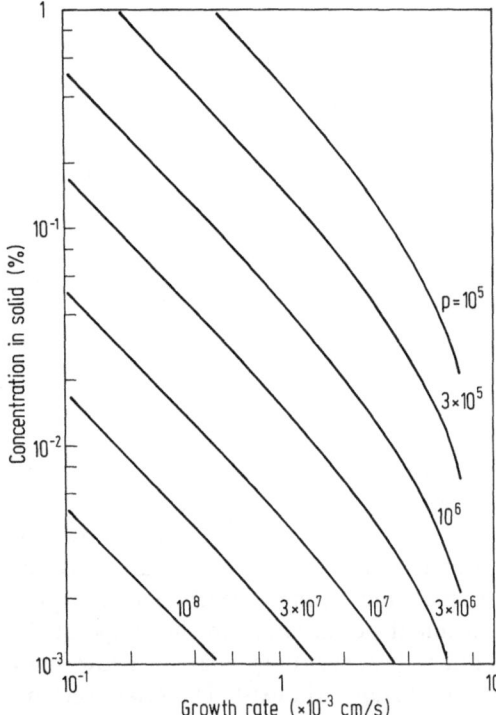

Fig. 9.3. Maximum solidus concentration achievable without constitutional
supercooling versus growth rate plotted
for several values of the parameter p =
− m (1 − k)/Dk. Temperature gradient
in the melt at the interface is taken to
be 50 K cm^{-1}. Material parameters
appropriate to germanium. [174]

on the onset of constitutional supercooling under the constraint of fixed solidus
concentration but stirring does suppress the onset under the condition of fixed
melt concentration of solute. These results can be expressed graphically as
maximum solidus concentration achievable at a given growth rate for a given
temperature gradient in the crystal as a function of a parameter:

$$p = - m(1 - k)/Dk \qquad\qquad 9.7$$

which depends on the choice of solute (Fig. 9.3).

9.3 Cellular Structures

The morphology of the cellular structure, (which can be revealed by rapidly
snatching the crystal from the melt), depends on the nature of the growth
mechanism at the interface. For materials which grow with a rough crystal-melt
interface with negligable kinetic supercooling for all crystal orientations, a close
packed honeycomb-like structure – which has the character of a close packed
hexagonal array in cubic materials (see Fig. 9.4) – is formed. This is in line with
the qualitative argument made in the last section concerning lateral diffusion
fields. This behaviour is seen in a wide range of low melting point metals and

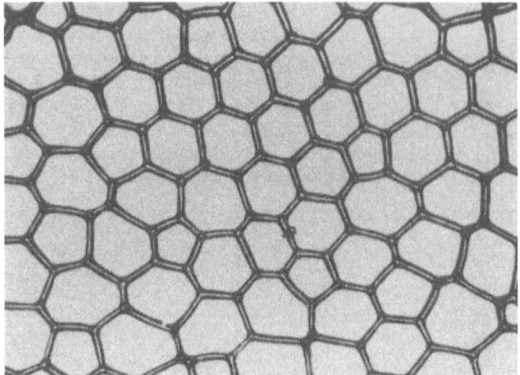

Fig. 9.4. Cellular structure produced by constitutional supercooling revealed by decanting the melt from a growing crystal of a lead-tin alloy (Mag × 52)

was extensively researched a number of years ago particularly by the Toronto school under Professor Chalmers (see Hurle [177] for a review).

If, however, the interface forms a low-index facet, then when growth is promoted on a convex isotherm normal to that direction, the pattern of the cellular structure becomes more complex and strongly dependent on crystal orientation. With such materials, an initial perturbation in the shape of the interface will grow in amplitude until it becomes tangential to a low-index faceting plane whereupon a nucleation difficulty will exist. The growth in this direction will cease temporarily and a facet will develop to a size which is sufficient to give the undercooling necessary to produce new nucleation. Thus the perturbed interface will develop into an array of inclined micro-facets. This is shown schematically in Fig. 9.5. The morphology of the cellular structure will therefore depend on the orientation of the major facets with respect to the growth direction. This is shown in Fig. 9.6 which is of the decanted interfaces of two identically grown crystals of gallium doped germanium, one grown on a [110] and the other on a [100] axis. The onset of this cellular structure is visible on the free cylindrical surface of the crystal as a set of "corrugations" corresponding to the intersection of the cell boundaries with the surface (Fig. 9.7a and b).

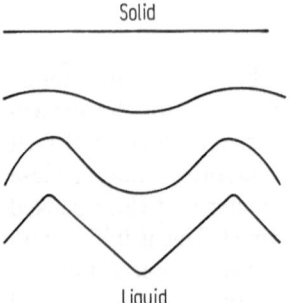

Fig. 9.5. Schematic representation of the stages of development of a faceted cellular interface

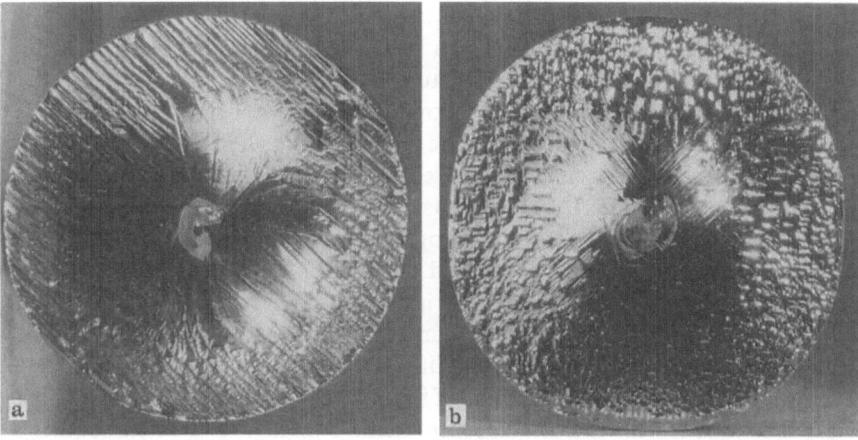

Fig. 9.6. Decanted interfaces of **a** a [110] and **b** a [100] oriented crystal of Ga-doped Ge showing cellular structure (Mag × 2.4)

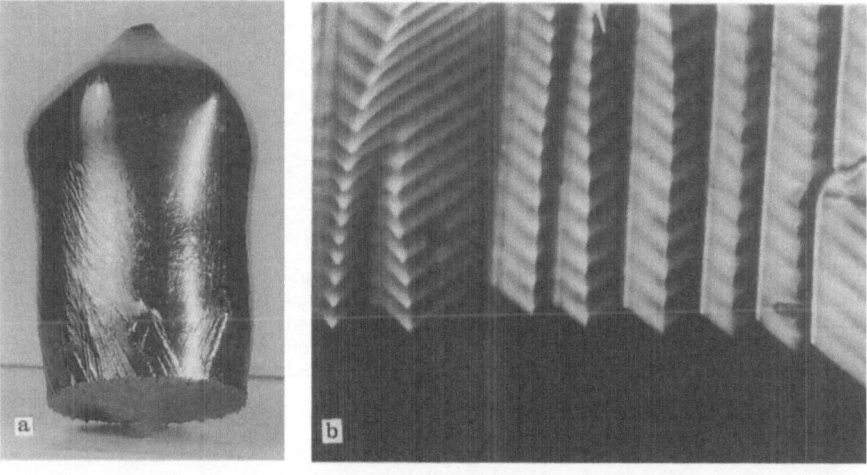

Fig. 9.7. Corrugations on the exterior surface of a ⟨110⟩ oriented Ga-doped Ge crystal. **a** Whole crystal (Mag × 1.2) and **b** magnified profile of the decanted interface. The herring bone structure is formed by the rotational growth striations (Mag × 80)

Thus far we have focused attention only on a single solute present during the growth of an elemental material. Constitutional supercooling can occur also during the growth of a compound from a non-congruent melt and indeed it is often a serious problem which is difficult to avoid, requiring extremely slow and steady growth rates. In general it precludes one from obtaining good quality single crystals of peritectically melting compounds by growth from the melt [137].

Growth from non-stoichiometric melts is here illustrated by the growth of indium antimonide from an indium-rich melt. Excess indium, rejected at the growing crystal-melt interface, gives rise to constitutional supercooling and the concommittant cellular structure as shown in Fig. 9.8. The decanted [100] interface is similar to the germanium one shown in Fig. 9.6b. However, a very small amount of radio-tellurium was added to the melt. This was preferentially incorporated onto the micro [111] facets (due to the facet effect described in Chap. 6) and is revealed on the autoradiograph shown in Fig. 9.8, proving that the cellular structure is indeed composed of small [111] facets.

Having shown that the cellular structure is composed of microfaceted segments, it is clear that a marked change in cellular morphology is to be expected when the macroscopic plane of the interface coincides with one of these faceting planes. In this case large areas of facet develop which are stable against cellular breakdown with the cell structure being confined to the neighbouring non-faceted areas of the crystal. This is well illustrated in Fig. 9.9 which shows

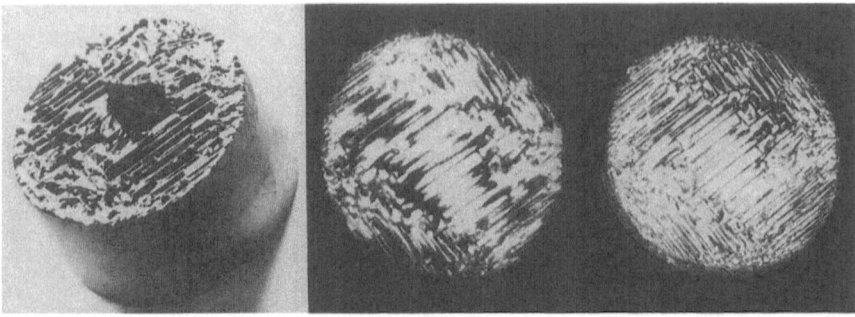

Fig. 9.8. Cellular structure of a [100] oriented single crystal of InSb produced by growing from an In-rich melt. (*right*) etched cross-section (*middle*) autoradiograph of the distribution of radio-tellurium dopant. Dopant is preferentially segregated to the micro-faceted regions of the growth interface (facet effect) thus demonstrating that the cellular structure is composed of microfacets. (*left*) macrograph of crystal snatched from the melt showing cellular interface. (Hurle et al. [243]) (Mag × 1)

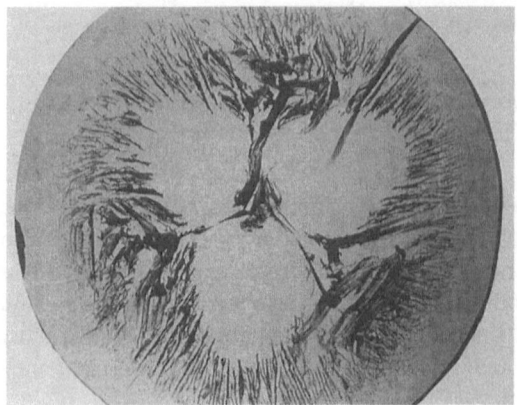

Fig. 9.9. Cellular structure in a [111] oriented single crystal of neodynium gallium garnet. [240] (Mag × 4.2)

the cellular interface of a pulled [111] axis crystal of neodynium gallium garnet with the central region dominated by three large {211} facets surrounded by regions of cellular structure.

Materials which form facets do not grow dendritically in the way that metals do because of the nucleation barrier which exists to propagation in these low-index faceting directions. However, the common diamond-cubic and zinc-blende structured semiconductors exhibit a curious form of pseudo-dendritic growth in which a lath-like structure, with [111] surfaces and containing two or more closely spaced twin planes, can propagate rapidly in the plane of the lath because the presence of these twin planes gives rise to a self-perpetuating re-entrancy thereby avoiding the nucleation problem (Wagner [111]). In the presence of large amounts of constitutional supercooling, such laminar-twin structures develop and the growth takes on a pseudo-dendritic form.

The above discussion illustrates the external manifestations of the cellular structure. If we now examine crystal sections by etching/optical microscopy and by X-ray topography, a wealth of fascinating detail can be revealed, principally because we can expose the interface morphology by delineating the growth striae as explained in Chap. 6. This is well illustrated in Fig. 9.10 which is of a [100] section normal to the growth axis of a gallium doped germanium crystal. The growth interface consists of an array of square and rectangular base pyramids having [111] faces. The regions separating neighbouring cells are the solute/rich cell boundaries. The growth striae delineate the trace of the four [111] planes in the [100] etched section at successive rotations of the crystal during growth. The straight lines, parallel and diagonal to the cell boundaries

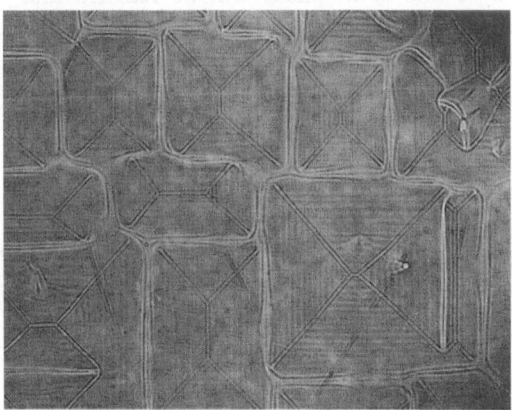

Fig. 9.10. [100] section of a Ga-doped Ge crystal normal to its growth axis etched to reveal striations delineating a cellular structure. Wavy lines defining approximately rectangular structures are the cell boundary grooves. Diagonal double lines (the doubling is an optical artifact resulting from taking the micrograph slightly out of focus to enhance the contrast of the striations) delineate the intersections of pairs of {111} microfacets. The terracing around the rectangular structures are rotational growth striations delineating successive layers of growth as the crystal is rotated (Mag × 42)

are the traces in the surface of the ridges formed by the intersection of adjacent pairs of [111] facets of an individual cell.

The rotation of the crystal imposes an overall pattern on the cellular structure as can be seen from the elongation of the cells on the [100] crystals shown in Fig. 9.6. The manner by which this occurs is evident from Fig. 9.11 which is a micrograph of an etched section normal to the growth axis of a ⟨110⟩ Ga-doped Ge single crystal sectioned near the point of onset of the cellular structure. Cells form initially near the crystal axis and the local turbulence which this induces is advected outwards giving the multi-start spiral distribution of solute shown in the figure by the enhanced etch rate. This pattern of solute distribution in the melt adjacent to the crystal at radial positions away from the axis markedly influences the pattern of cellular formation in these regions. Specifically, for the case of ⟨110⟩ axis growth shown, the long axes of the cells tend to be terminated at the spiral arm positions so that cells are elongated when their long axis is parallel to the spiral arms and truncated when orthogonal to them.

By studying longitudinal sections of the grown crystal one can learn about the evolution of the cellular structure. Fig. 9.12 is a micrograph of a (110) longitudinal section of a ⟨110⟩ axis gallium doped germanium crystal. Growth striae delineate the profile of the cellular morphology which is composed of pairs of {111} planes inclined at 35° to the plane normal to the growth axis.

These cell boundary grooves (the narrow channels between the cellular projections) become progressively richer in solute (for a solute with segregation coefficient less than unity) and become deeper because growth is impeded by the solute enrichment. Ultimately, as the region at the root of the cell boundary cools, a phase boundary with a second solid phase will be reached and a precipitate will be formed, often by a eutectic reaction. Alternatively, if the freezing range is very wide, the deep channels become unstable as the cell boundary concentration increases and spheroidisation occurs leaving liquid droplets entrapped within the crystal (Fig. 9.13). These droplets experience the

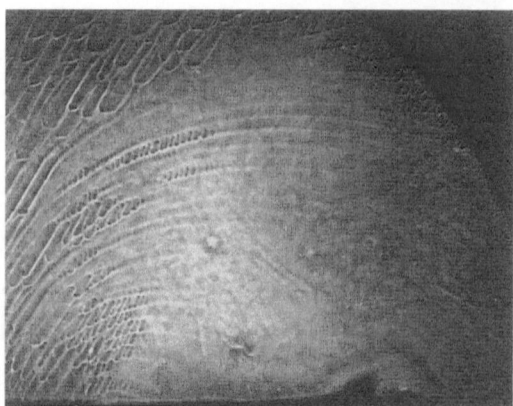

Fig. 9.11. {110} section normal to the growth axis of a Ga-doped Ge single crystal showing spiral pattern of solute distribution caused by turbulence generated at cellular projections formed in the vicinity of the crystal axis (the bottom left hand corner of the photomicrograph) (Mag × 6)

Fig. 9.12. (110) longtitudinal section of a ⟨110⟩ oriented crystal of Ga-doped Ge etched to reveal striations delineating a cellular structure. Single vertical lines mark the trace of the cell peaks; multiple vertical lines are the trace of the cell boundaries. The herring bone pattern is produced by striations generated by crystal rotation. The growth direction is vertically downward (Mag × 60)

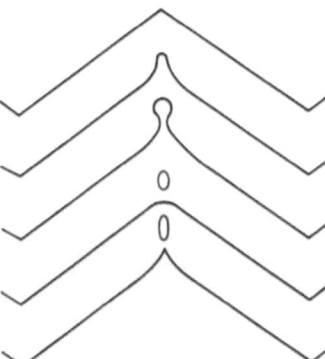

Fig. 9.13. Schematic representation of droplet entrapment in a cell boundary groove. Growth direction is ⟨110⟩

imposed temperature gradient in the crystal and dissolve on the hotter side and solidify on their cooler side, i.e. they "climb" through the crystal following the main crystal-melt interface by a process known as temperature gradient zone melting (TGZM) [178] leaving behind highly solute-rich material in the form of "solute trails". This can be seen in Fig. 9.14. When the crystal is finally cooled to room temperature, the droplets solidify as second phase inclusions. For some dopants, the concentration of solute in the solute trail is sufficiently high for it to be revealed by X-ray absorption topography as shown in Fig. 9.15.

The forms which these cell boundaries and solute trails take depend on the crystal orientation. If facets are symmetrically disposed to the growth direction, then the cell boundaries will be aligned parallel to that direction and the migrating droplets will be confined to the cell boundary regions (as in Fig. 9.13). However, if the facets are asymmetrically inclined to the growth axis, the cell

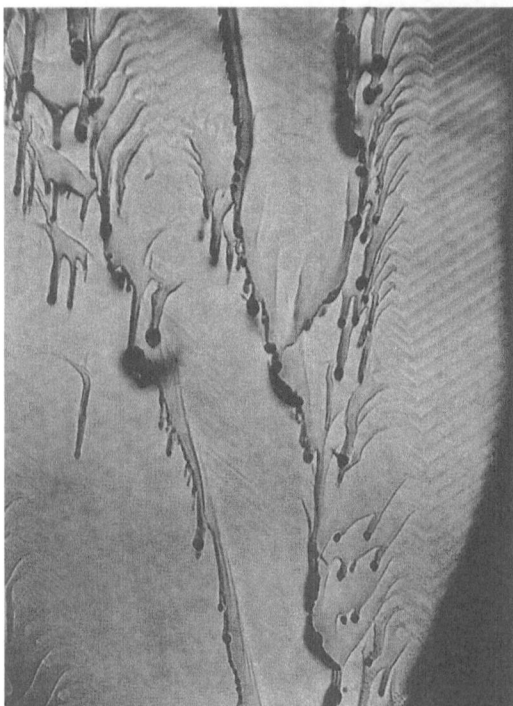

Fig. 9.14. Solute trails in the longitudinal section of a $\langle 110 \rangle$ oriented single crystal of Ga-doped Ge revealed by etching (Mag × 74)

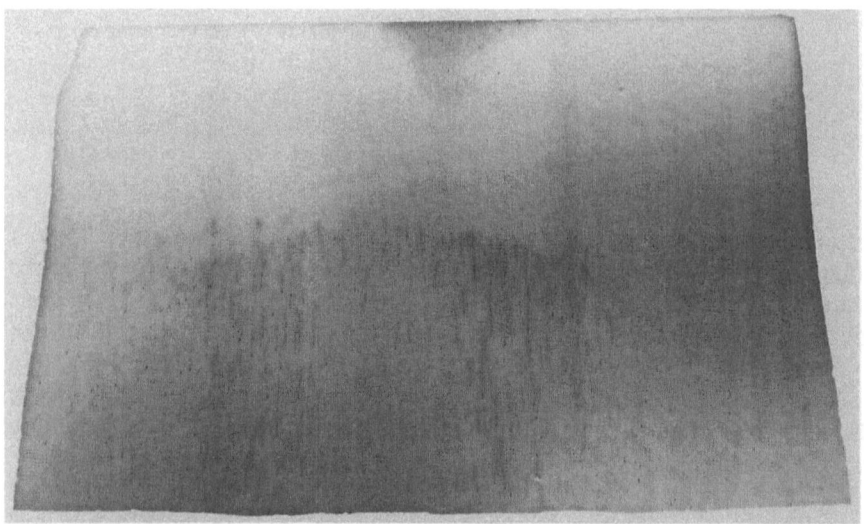

Fig. 9.15. Solute trails in a Sn-doped Ge single crystal revealed by X-ray absorption topography. Dark contrast corresponds to Sn-rich crystal. The dark 'dots' are the solidified droplets and probably contain Sn-Ge eutectic. (Bardsley et al. [244]) (Mag × 5.2)

boundaries will also be inclined to it but the liquid droplets, which climb parallel to the temperature gradient, will move along the growth axis and will therefore not be confined to the cell boundaries. The latter situation is depicted in Fig. 9.16.

The thermodynamics of the system also has an important bearing on the form of the micro-inhomogeneity. If the solute-melt system has only a small freezing range, then only a little migration of the liquid droplets will be possible because a solidus composition will be reached quickly as the crystal cools. However, if, as is the case of gallium doped germanium illustrated in several of the figures, there is a very extended liquid range, then extensive droplet migration occurs.

If the solute raises the freezing point (i.e. the segregation coefficient is greater than unity) then the solute is preferentially segregated to the peaks of the cell and the cell boundaries are depleted in solute. Since this depletion can only occur down to zero percent solute, the depth of the cell boundaries in this case is limited.

The sign of the volume change on solidification of the material in the cell boundary groove has an important influence. The common semiconductors expand on solidification and therefore solidification of the cell boundary material and of the migrating droplets places the surrounding crystal in compression. If, on the other hand, as is common with oxide crystal growth, there is volume contraction on solidification, then freezing of the cell boundary liquid can give rise to long tubular voids in the crystal as is exemplified in Fig. 9.17 which is of a calcium tungstate crystal.

The above paragraphs have briefly illustrated the fact that very gross chemical inhomogeneities can be incorporated into crystals in the presence of a zone of constitutional supercooling in the melt from which the crystal is formed. Not surprisingly the stresses which these inhomogeneities generate can, at the high temperatures involved, exceed the yield stress and dislocation arrays are

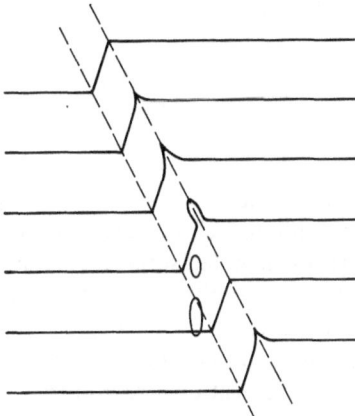

Fig. 9.16. Schematic illustration of solute trails that have climbed away from the cell boundary grooves; (⟨111⟩ growth axis crystal)

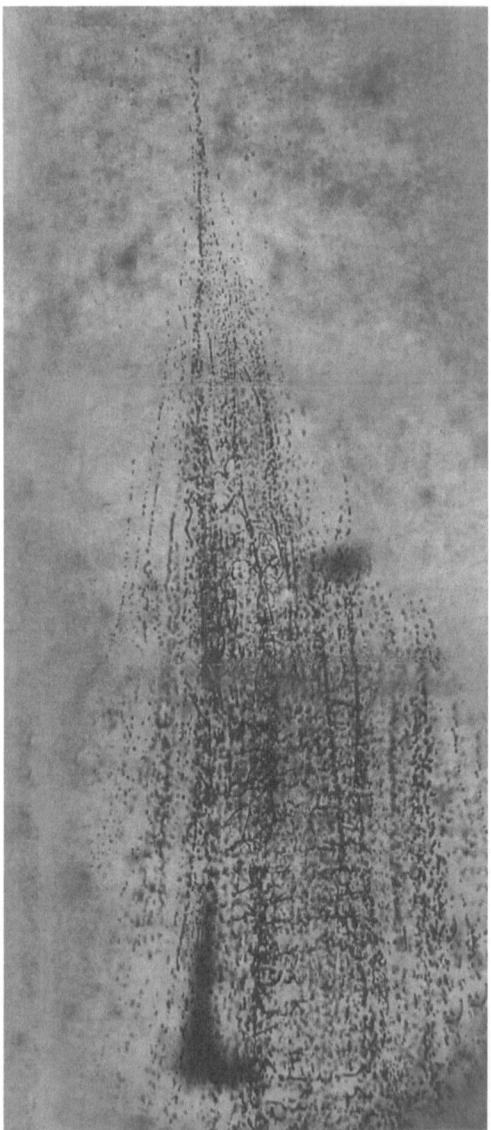

Fig. 9.17. Tubular voids in a calcium tungstate crystal containing a cellular structure; longtitudinal section. Growth direction is vertically downward. [222] (Mag × 12)

generated in the crystal. Such arrays are readily revealed by chemical etching and/or by X-ray topography. In the case of germanium, etching {111} surfaces reveals dislocation etch pits whilst etching all other orientated surfaces reveals a pattern of microsegregation which delineates the topography of the interface.

Dislocation arrays form in both the cell boundaries and around the solute trails and their solidified liquid droplets. For a segregation coefficient less than unity, the cell boundaries are delineated by massive dislocation arrays as seen in Fig. 9.18.

Fig. 9.18. Massive dislocation arrays, delineated by dislocation etching, in the cell boundaries of a ⟨110⟩ oriented single crystal of Ga-doped Ge. Lines delineating the cell peaks can be seen midway between the dislocated cell boundaries. Circular patches of lower dislocation density visible in some of the cell boundaries show where solute trails have climbed through, re-dissolving the highly dislocated material and recrystallising it at a lower temperature where the lattice is stronger (Mag × 37)

Dislocation arrays associated with solute trails can take a variety of forms. For solutes with a segregation coefficient not much less than unity, the migration of the droplet can result in the loss of solute so that the droplet shrinks as it climbs and peters out in a short distance. If however a second phase precipitate forms, then a pattern of octahedral slip is evident, as in Fig. 9.19a.

The formation of these arrays of slip dislocations represents a dramatic deterioration of the structural perfection of the crystal as can be seen from Fig. 9.20 which is of a longitudinal section of a [110] gallium-doped germanium crystal.

Growth on a ⟨111⟩ axis with a convex-to-the-melt interface produces a macro-faceted structure with a large central faceted region within which good

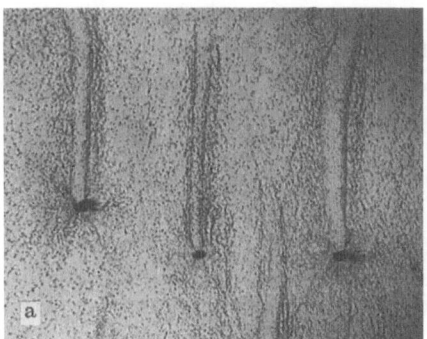 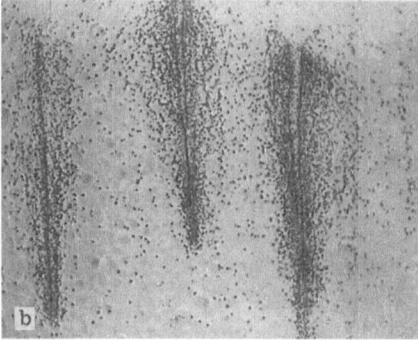

Fig. 9.19. Dislocation arrays associated with solute trails. **a** High solute concentration; note evidence of octahedral slip where the droplet solidifies; **b** light doping; solute trail peters out due to exhaustion of solute (Mag × 34)

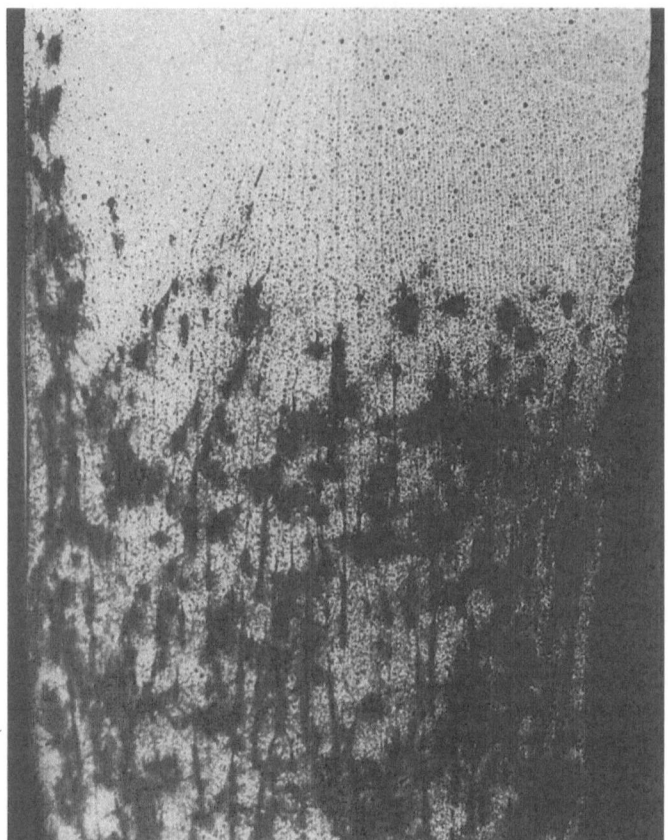

Fig. 9.20. Longitudinal section of a ⟨110⟩ oriented lightly Ga-doped Ge crystal showing evolution of dislocation structure as cellular structure develops. Cellular structure first forms approximately at the top of the micrograph. The vertical lines in the top half of the micrograph are the delineation of the cell boundaries (Mag × 4)

crystallinity is preserved. Re-entrancies formed at the macro-steps trap solute and generate high dislocation densities as shown in Fig. 9.21 which is of a slowly rotated boron-doped Si single crystal.

Growth of crystals on a [111] axis with a concave-to-the-melt interface produces a set of peripheral {111} macrosteps whose motion is governed by the crystal rotation. This is shown in Fig. 9.22a which is of the dislocation pattern of such a crystal of gallium doped germanium. Figure 9.22b shows an auto-radiograph of the Ga-distribution in the same crystal. Note the similarity of the autoradiograph of the gallium distribution with the etch pit distribution which reveals dislocations. Essentially similar behaviour occurs with oxide crystals. This is illustrated in Fig. 9.23 which is of the decanted interface of a (001) interface of a calcium tungstate crystal and Fig. 9.24 shows a longitudinal section through the same crystal.

Fig. 9.21. Section normal to the ⟨111⟩ growth axis of an boron-doped silicon single crystal etched to reveal dislocation arrays formed at the macrosteps of the cellular structure produced on a convex crystal-melt interface (Mag × 2.5)

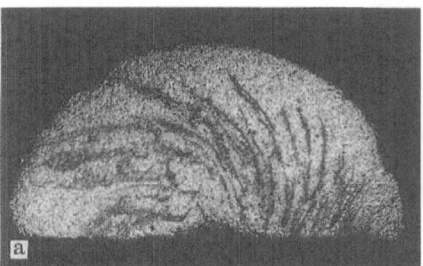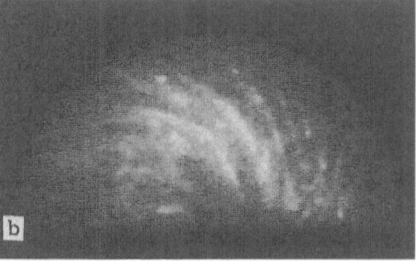

Fig. 9.22. a Spiral arrays of dislocations due to propagation of the cellular structure on a concave interface of a ⟨111⟩ oriented Ga-doped Ge single crystal; **b** autoradiograph of the Ga-distribution in the same crystal section (Mag × 2.5)

Fig. 9.23. Decanted interface of a ⟨001⟩ oriented single crystal of calcium tungstate showing cellular structure. (Bardsley et al. [245]) (Mag × 60)

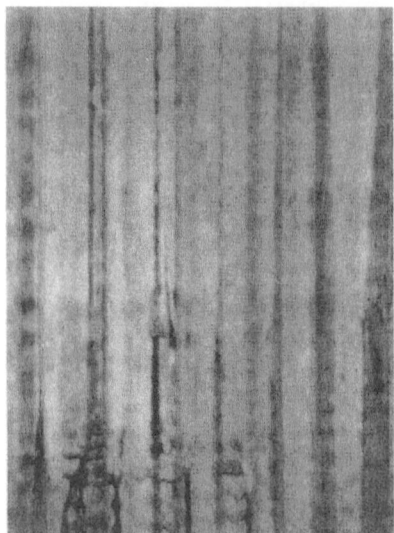

Fig. 9.24. Longtitudinal section of the crystal shown in Fig. 9.23 (Mag × 45)

9.4 Linear Stability Theory

Having given a semi-quantitative explanation for the formation of a cellular structure and described its occurence and manifestations in pulled crystals, we now step back and provide a more rigorous analysis of the morphological stability of the growing interface. This approach was pioneered in a classic paper by Mullins and Sekerka [179].

Consider an initial state of a crystallising system in which a planar interface, located on the plane $z = 0$, is solidifying at a constant velocity v_p. We first solve for the temperature and solute distribution in this system and then consider an infinitesimal perturbation in the shape of the crystal-melt interface which is expressed as a Fourier component in the Y direction. The question is then asked as to whether this perturbation grows or decays and the conditions for the critical state for which the perturbation is neutrally stable are defined. The governing equations of heat and mass transfer expressed in a reference frame located on the unperturbed crystal-melt interface, excluding convective effects, are:

$$\kappa_s \nabla^2 T_s + v_p \partial T_s / \partial z = \partial T_s / \partial t \qquad 9.8$$

$$\kappa_L \nabla^2 T_L + v_p \partial T_L / \partial z = \partial T_L / \partial t \qquad 9.9$$

$$D \nabla^2 C + v_p \partial C / \partial z = \partial C / \partial t \qquad 9.10$$

where $\kappa_{s,L}$ are the thermal diffusivities of solid and liquid respectively and D is the solute diffusion coefficient in the liquid. Diffusion in the solid is ignored.

The boundary conditions on the solute field are that, remote from the crystal, the concentration has a fixed value C_∞ and at the crystal- melt interface a flux balance condition is satisfied. These conditions expressed mathematically are:

$$C = C_\infty \quad \text{at } z = \infty \qquad\qquad 9.11$$

and

$$D[\partial C/\partial z]_0 + v_p(1 - k)\,C(0) = 0 \quad \text{at } z = 0 \qquad\qquad 9.12$$

where k is the interface segregation coefficient of the solute.

The thermal and solute fields are coupled by the condition of thermodynamic equilibrium applied at the crystal-melt interface, viz:

$$T(0) - T_M = mC(0) \qquad\qquad 9.13$$

where m is the slope of the liquidus of the binary alloy and T_M is the melting point of the major constituent. Kinetic effects at the crystallising interface are neglected.

In this coordinate system moving with the unperturbed interface the system is at steady state.

Further, the thermal Peclet number is very small, viz:

$$v_p d/\kappa_{S,L} \ll 1$$

where d is some characteristic dimension of the system. We can therefore omit the advection terms in Eqs. 9.8 and 9.9 so that we have, in the unperturbed state:

$$d^2 T_s/dz^2 = d^2 T_L/dz^2 = 0 \qquad\qquad 9.14$$

$$d^2 C/dz^2 + (v_p/D)\cdot dC/dz = 0 \qquad\qquad 9.15$$

The solution of Eq. 9.15 subject to the boundary conditions is:

$$C = C_\infty - (DG_c/v_p)\cdot\exp(-v_p z/D) \qquad\qquad 9.16$$

where

$$G_c = -(1 - k)\,v_p C_\infty/kD \qquad\qquad 9.17$$

is the concentration gradient at the interface. G_L is the temperature gradient in the melt obtained by integrating Eq. 9.14.

The temperature gradient in the solid is then simply:

$$G_s = (k_L G_L - Lv_p)/k_s \qquad\qquad 9.18$$

where L is the latent heat per unit volume.

We now examine the stability of this system to a disturbance of the form:

$$\Phi = \Phi(z)\,\exp[i(k_x x + k_y y) + pt] \qquad\qquad 9.19$$

where

$$\Phi = \bar{T}_S(z),\ \bar{T}_L(z),\ \bar{C}(z) \qquad\qquad 9.20$$

are the amplitudes of the disturbances of the thermal fields and solute field respectively. The interface will be perturbed from its initial planar state ($z = \phi = 0$) to:

$$z = \phi = \delta exp\ [i(k_x x + k_y y) + pt] \qquad 9.21$$

where δ is the (small) amplitude of a general Fourier component of the disturbance.

Substituting Eq. 9.19 into the working equations, Eqs. 9.8–9.10 and setting the thermal Peclet number to zero yields the following disturbance equations:

$$[D(\partial^2/\partial z^2 - \omega^2) + v_p \partial/\partial z - p]\bar{C} = 0 \qquad 9.22$$

$$(\partial^2/\partial z^2 - \omega^2 - p/k_s)\ T_S = (\partial^2/\partial z^2 - \omega^2 - p/k_L)\ \bar{T}_L = 0 \qquad 9.23$$

were

$$\omega^2 = k_x^2 + k_y^2$$

The thermal boundary conditions are now:

$$\bar{T}_s - \bar{T}_L + (G_s - G_L)\ \delta = 0 \qquad 9.24$$

$$k_s(\partial \bar{T}_s/\partial z - v_p G_s/\kappa_s \cdot \delta) - k_L(\partial T_L/\partial z - v_p G_L/\kappa_L \cdot \delta) - \mathscr{L}p\delta = 0 \qquad 9.25$$

where

$$\mathscr{L} = L/\rho C_p$$

The perturbed solute boundary condition is:

$$[D\partial/\partial z + v_p(1 - k)]\bar{C} - [kv_p G_c - (1 - k)\ C_\infty/k \cdot p]\delta = 0 \qquad 9.26$$

The local equilibrium condition becomes:

$$\bar{T}_L - m\bar{C} - [mG_c - G_L - T_M\Gamma\omega^2]\delta = 0 \qquad 9.27$$

where $\Gamma = \sigma_{SL}/L$ is the capillary constant, σ_{SL} being the crystal-melt interfacial energy. The set of equations, Eqs. 9.24, 9.26 and 9.27 define the stability of the system to first order.

These equations have solutions of the form:

$$\bar{T}_S = A_S exp(\omega z) \qquad 9.28$$

$$T_L = A_L exp(-\omega z) \qquad 9.29$$

$$\bar{C} = A_C exp(-\omega^* z) \qquad 9.30$$

where

$$\omega^* = -v_p/2D + [(v_p/2D)^2 + \omega^2]^{1/2} \qquad 9.31$$

Defining a Sekerka number:

$$S = mG_c/G \qquad 9.32$$

and a non-dimensional capillary parameter

$$A = T_M \Gamma v_p^2 / D^2 G \qquad\qquad 9.33$$

where

$$G = (k_S G_S + k_L G_L) / (k_S + k_L) \qquad\qquad 9.34$$

we obtain the condition for neutral stability of the disturbance:

$$S = \{2k - 1 + [1 + 4(a^2 + \sigma)]^{1/2}\}\{-1 + [1 + 4(a^2 + \sigma) - 2\sigma]^{1/2}\}^{-1}$$

$$\times [1 + Aa^2 + Lv_p\sigma/aG(k_S + k_L)] \qquad\qquad 9.35$$

where $a = D\omega/v_p$ and $\sigma = Dp/v_p^2$.

This is plotted in Fig. 9.25 for $A = 1$ for several values of the segregation coefficient. The critical value of the spatial wave number ($a = a_c$) corresponding to the minimum value of $S = S_c$ is obtained by setting:

$$dS/da = 0 \qquad \text{at } a = a_c \qquad\qquad 9.36$$

The critical wave number (a_c) is therefore the real root of:

$$A[(1 + 4a_c^2)^{1/2} - 1]^2[(1 + 4a_c^2)^{1/2} + k] = 4k \qquad\qquad 9.37$$

We can compare this result with the simple constitutional supercooling criterion derived in Sect. 9.2. The two criteria are identical if $S_c = 1$ and $G = G_L$. The latter condition is satisfied if the thermal conductivities of the two phases are equal and the latent heat is zero. By replacing G_L by G, one is taking account

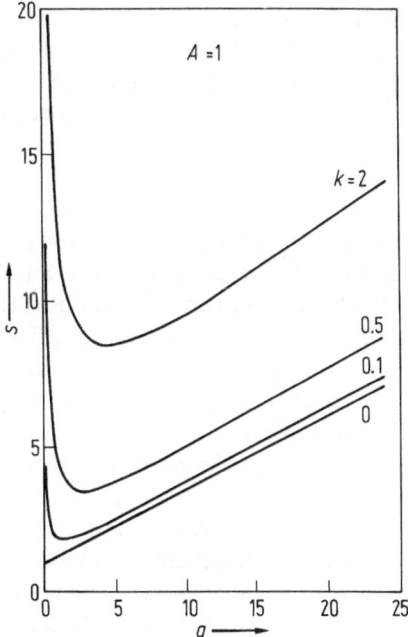

Fig. 9.25. Marginal stability curves: stability parameter versus wave number a plotted for various values of the segregation coefficient. Capillary parameter $A = 1$ (see text)

of the redistribution of the heat flow as a result of the deformation of the interface. The condition $S_c = 1$ is approached in two limits: firstly, if the Gibbs-Thomson effect is negligible ($A \to 0$), then the critical condition becomes:

$$S_c = 1 + 3(Ak^2/4)^{1/2} \qquad\qquad 9.38$$

with $a_c = (k/2A)^{1/3}$.

The other limiting case is of zero segregation coefficient (i.e. where the solute is insoluble in the crystal) whereupon the critical Sekerka number is:

$$S_c = 1 + 2(kA)^{1/2} \qquad\qquad 9.39$$

Under practical experimental conditions it is not readily possible to differentiate between the two theories except under conditions of extremely high growth velocity (such as are achieved by laser annealing) when the Gibbs-Thomson effect can become very important.

The above analysis neglects any convection in the melt and, to apply this theory effectively to Czochralski growth, account must be taken of the effects of melt flow. This is a complex problem which is still receiving active study [180], [181].

The simplest, but least rigorous, method of taking account of the flow is to utilise a boundary layer (δ) concept. This was first done by Birman [182] and by Coriell et al. [183] and subsequently refined by Hennenberg and co-workers [184, 246]. The analysis is considerably more complicated and is not reproduced here. In the limit δ becomes infinity, it reduces to the Mullins and Sekerka result (Eq. 9.35 with 9.37).

The effect of forced parallel flows over the interface has been considered by several workers [185], [186]. However, these are not very relevant to the case of Czochralski growth since the swirling von Karman type flow generated by the rotating disc is certainly not a parallel flow. This problem has been tackled by Brattkus and Davis [187]. The essential mathematical problem is that such non-parallel flows cannot be analysed using standard normal modes. To overcome this difficulty, the authors restricted their attention to slowly rotating crystals where the thickness of the viscous boundary layer was much greater than that of the solute boundary layer. They found that long wavelength disturbances of the interface were particularly sensitive to the non-parallel nature of the flow. Such long waves, which are stable by the Mullins and Sekerka criterion in the absence of flow, were shown to be highly unstable in the presence of the flow. The authors term this effect "flow-induced morphological instability". They found that the flow was destabilising to the point that any degree of constitutional supercooling would result in a morphological instability.

Whilst it may not be true for all crystal growth configurations, it would appear therefore that, for the Czochralski case at least, the criterion of constitutional supercooling is a perfectly adequate one for the description of morphological instability. Brattkus and Davis showed that, as the Chalmers number $M = m\,G_c/G_l$ increased past unity, the system became unstable to travelling waves moving against the flow and toward the axis of rotation. This instability is due to the redistribution of solute by both the horizontal and vertical components of the flow velocity. The vertical velocity directed towards the crystal destabilises the interface by compressing the thickness of the solutal boundary layer and thereby increasing the solute gradient at the interface. The horizontal flow, which is dependent on the radial position, increases in strength with distance from the origin and this radial gradient of radial velocity produces a variation in the ability of the flow to mix the solute and further destabilises the interface by creating horizontal gradients of solute. It is these horizontal gradients which are the driving force propagating the travelling waves against the flow. Clear experimental evidence for this phenomena has not been obtained except perhaps for an interesting photograph from a paper by Singh et al. [188] which shows a modulation in the rotational striations in a germanium crystal near the onset of constitutional supercooling which may well represent the travelling waves described above. This is reproduced in Fig. 9.26. Brattkus and Davis [187] suggest however that these long wave modes may in fact be the rotational striations themselves. However, this suggestion does not appear to be compatible with the observation that rotational striations are present even under conditions well away from those corresponding to constitutional supercooling.

Finally, we mention the interesting idea explored by Wheeler [189] that one might influence the onset of the morphological instability by periodically modulating the growth rate. Were it possible to suppress the onset of morphological instabilities significantly by such a method, then the prospect is offered of growing uniform alloy single crystals by pulling from the melt simply by modulating the pulling speed at an appropriate frequency. However, a detailed analysis by Wheeler shows this idea to be sound in principle but the calculated degree of stabilisation under realistic growth parameters is insufficiently large to

20 μm

Fig. 9.26. Etched longtitudinal section of a Ga-doped Ge single crystal showing travelling waves on the interface. The *arrow* indicates the growth direction. (Singh et al. [188])

give rise to a worthwhile technique. The required amplitude of the modulation to obtain any observable effect must be sufficient to generate melt-back during part of the oscillation cycle.

Merchant and Davis [247] have considered plane stagnation-point flow but with the flow at infinity modulated periodically. They deduced that low frequency modulation stabilised the interface against flow-induced morphological instability whilst a high frequency promoted the instability.

10 The Cooling Crystal

10.1 Introduction

The properties of a Czochralski-grown crystal depend not only on the conditions prevailing at the point of growth (e.g. the solute distribution resulting from segregation at a planar or cellular interface fed from a melt whose solute distribution is determined by the complex flow patterns in that melt) – but also on the processes which occur as that crystal cools from the melting point down to room temperature. Some of these changes result from attempts by the crystal to remain in or near to local thermodynamic equilibrium whilst others are non-equilibrium effects – such as the introduction of dislocations into the crystal.

In this chapter, we consider examples of both of these topics with respect to a single material, gallium arsenide. The relatively small critical resolved shear stress possessed by gallium arsenide means that this critical parameter is exceeded by the thermal stresses generated in the cooling gallium arsenide crystal with the result that dislocations are generated. (This can be avoided in the case of silicon and indeed the commercial production of this material is to a specification which demands that there are no extended dislocations in the crystal).

A single dislocation threading the active layer of a gallium arsenide laser causes that laser to "die" within a relatively short operational lifetime [220]. Additionally, the uniformity of the threshold voltage of an array of field effect transistors on a wafer has been shown to depend on the proximity of the transistor to the nearest dislocation [221]. These deleterious effects of dislocations in the material have led to an extensive study, both experimentally and by numerical simulation, of the origin of stress-induced dislocations in gallium arsenide. In contrast, for many oxide materials the Burger's vector of a dislocation is so large that dislocation generation is energetically unfavourable, deformation twinning and fracture occurring preferentially. In such materials the problem is not how to avoid dislocation generation but rather how to avoid the crystal falling apart as it cools – an even more serious state of affairs!

Thermodynamic related effects on the other hand are common to all materials where kinetic effects are not too seriously rate-limiting. Thus, as the crystal cools, it can pass through a region of parameter space which crosses a phase boundary leading to precipitation of some second phase. An example of this is shown in Fig. 10.1 where the retrograde solidus of gallium arsenide results in the precipitation of hexagonal arsenic on dislocations.

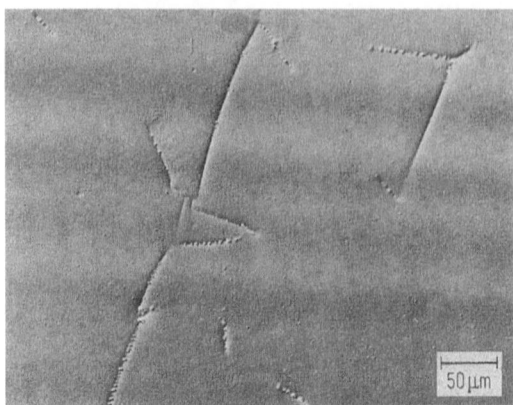

Fig. 10.1. Optical Nomarski interference micrograph of an AB etched section of a Zn-doped GaAs single crystal showing arsenic precipitates on dislocations. (Cullis et al. [248])

In addition to considerations of the formation of a second phase is the question of the concentration of native point defects in solution in the crystal at any given temperature. As the crystal cools, elemental point defects such as vacancies and interstitial atoms which are grown-in at the melting point, aggregate to form more complex native defects. In the case of semiconductor materials this situation is particularly complicated because these defects can be, and usually are, charged, their charge state depending on the position of the Fermi level in the material which itself will change as the crystal cools.

From the above it can be seen that the control of the conditions during cooling of the crystal are quite as important as controlling the conditions of growth. To this end after-heaters [222] are frequently employed, especially in the case of oxide crystal growth, principally in order to control the thermal stresses. In the case of semiconductors, post-growth heat treatment schedules are often devised to establish the required point defect concentration and configurations.

10.2 Stress Distribution and Dislocation Generation in Gallium Arsenide and Indium Phosphide

It is well established that crystallographic glide, induced by the excessive thermal stresses during cooling of the crystal, is the major source of dislocations in as-grown ingots of the III/V compounds [190]. Because of radial heat loss from the cooling crystal, its periphery in any plane normal to its axis, is cooler than a point on the axis. The finite thermal expansion of the material therefore results in the exterior of the crystal being in tension with the interior in compression. These two regions are divided by a neutral cylindrical surface at some radial position. This simple picture is complicated by the axial temperature gradient in the crystal and it can be readily shown that this contributes to the stress by an amount related to the axial derivative of the axial temperature gradient (i.e. to d^2T/dz^2, where z is the axial coordinate [191]). In germanium and silicon these thermal stress-induced dislocations tend to remain on their slip planes producing a pattern, for growth on a $\langle 111 \rangle$ axis, like that shown in Fig. 10.2.

In GaAs on the other hand these stress- induced dislocations appear to climb out of their slip planes by the condensation of native point defects in the crystal to form a polygonised structure which is very characteristic of this class of material grown by the LEC technique (Fig. 10.3). The polygonisation appears to set in only above some critical dislocation density. If only small diameter crystals are grown, then it is possible to obtain zero dislocation density. If such a small crystal is steadily increased in diameter then, at a critical diameter which is dependent on the growth environment, dislocations are generated, initially confined to their glide planes [192]. Equally, if indium is added to the melt so as to produce an alloy single crystal containing around one or more percent

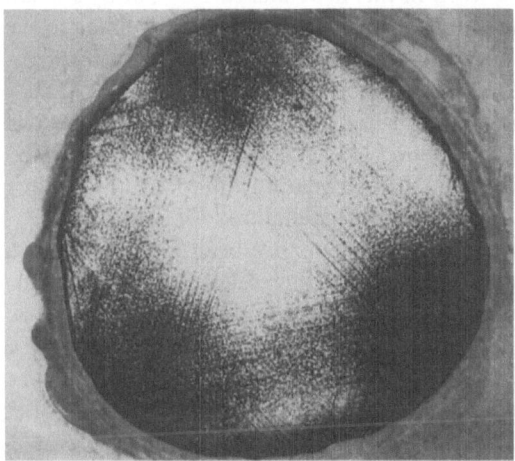

Fig. 10.2. Dislocations, revealed by etching, lying on {111} slip planes in a $\langle 111 \rangle$ oriented germanium single crystal (Mag × 2.2)

Fig. 10.3. Cellular dislocation structure in a semi-insulating GaAs crystal generated by climb and glide of dislocations formed by plastic flow to relieve elastic stress in the cooling crystal (Mag × 16)

indium arsenide, then it is found that dislocation generation in such a crystal is dramatically suppressed [193] such that, with care, nearly dislocation-free crystals of 5 cm, 7.5 cm and even 10 cm diameter can be grown. Initially it was thought that this arose because addition of the indium "hardened" the lattice, i.e. raised its critical resolved shear stress. However, measurements by Tabache et al. [194] showed that the increase in critical resolved shear stress which resulted from alloying with indium was far from sufficient to account for the dramatic reduction in dislocation density so that, whatever the mechanism involved with the indium addition, simple solution hardening does not explain it. The authors in fact suggest that indium doping inhibits dislocation formation by modifying the thermodynamic equilibrium of native defects in the crystal. This issue is at present unresolved.

In order to predict the dislocation density in a grown crystal we need to be able to predict the temperature distribution at all stages of its cooling and to have some constituitive relationship between the temperature field and the density of dislocations. Pioneering work in this field was carried out by Jordan et al. [195], who supposed that the dislocation density was linearly related to the amount by which the maximum thermal stress, experienced at a point through-out the cooling cycle, exceeded the critical resolved shear stress (CRSS). They were able to show, using this simple empirical postulate, that the pattern of dislocations observed in as-grown [100] and [111] crystals broadly matched the pattern of excess thermal stress calculated from a heat flow model of the cooling crystal. A similar approach was adopted by Milvidskii and Golovin [196].

Extensive modelling based on Jordan's concepts has been carried out by a large number of authors [197–204]. Their simulations differ principally in the sophistication employed to model the thermal distribution. Thus analytical models (which essentially apply for infinitely long crystals) have been developed by Szabo [205] and by Kobayashi and Iwaki [206]. At the other end of the spectrum Atherton et al. [84] and Crowley et al. [90] have used radiation optics to realistically model the radiative transfer environment within the growth chamber in order to obtain an accurate numerical model of the temperature field. However, a crucial factor is the heat transfer processes which occur within the boric oxide encapsulant. Since the relevant thermal and optical parameters are not well known for this material, the modelling is only as good as the parameters are truly representative of the boric oxide. Motakef and Witt [198] have shown that the maximum stresses are strongly affected by the thermal transparency and thickness of the liquid encapsulant. For a fully transparent medium the stress levels are essentially the same as they would have been without the presence of the encapsulant, whereas for a wholly opaque one, there is a reduction of thermal stress in the crystal with increasing encapsulant thickness. However, significant stresses due to radial temperature gradients are generated at the point of emergence of the crystal from the encapsulant where the heat losses increase abruptly. Meduoye and co-workers [199–201] concen-trated on the dislocation generation in the conical portion of the crystal since they had shown that the stress is greatest here and is the point at which dislocations are first introduced into the crystal. Unless dislocation introduction at this point can be avoided, there is little chance of obtaining low dislocation density in the subsequently-grown part of the crystal.

It was noted in Chap. 5 that semi-transparency of the crystal produced an increased radial temperature gradient [110]. This exacerbates the dislocation generation problem and may account for the very high dislocation densities commonly seen near the crystal surface. However this effect has not, up to the present time, been incorporated into the stress-field models.

Medouye et al. [199–201] showed that the conditions for minimising the thermal stress are different in the two cases of cone development and cylindrical growth. The ideal situation for both cases is a purely axial temperature gradient. Since the axial gradient does not appear to vary significantly with axial position, it is only radial stresses which are important and these are zero for zero radial temperature gradient. Planar isotherms are achieved for the cylindrical portion of growth when there is zero radial heat loss at the cylindrical surface of the crystal, i.e. when the boric oxide provides an insulating blanket. In the cone on the other hand, maintenance of a planar isotherm requires a matched heat loss into the boric oxide at the conical periphery of the crystal. Meduoye et al. show that this desirable state can be approached by judicious choice of growth parameters particularly with attention to the quantity and temperature of the encapsulant. The depth of the oxide is a crucial factor; more important than its thermal conductivity. High stress concentrations occur at the gallium arsenide-ambient gas-encapsulant triple junction which can only be avoided by matching the radiative heat losses from the gallium arsenide just above this line to those produced by conduction just below it. This suggests the use of a combination of reflectors and adjustable after-heaters.

Similar conclusions have been drawn by Motakef [202] who considered the effects of the encapsulant properties and the environmental temperature distribution on the growth of cylindrical crystals. By increasing the depth of the boric oxide one removes this region of high stress to a position of lower temperature where the CRSS is greater and therefore the crystal is more able to withstand that stress. However, this strategy brings its own problems; principally of an increased difficulty in diameter control. For a given thermal environment and encapsulant depth, the stresses are a function of the angle of the cone and in general there will be some critical angle, which is likely to be quite small, for which the dislocation density, integrated over the crystal volume, is minimised. However, small cone angles imply large cone lengths before the desired diameter is reached and this brings significant economic disbenefit.

The CRSS is taken to be the lower yield stress of the material considered to be that value of the stress which initiates dislocation generation. (Numerical values are obtained from compression testing). The excess stress is obtained by subtracting this from some weighted average of the components of the stress tensor such as the Von Mises stress. Jordan et al. [197], for example, summed the absolute values of the twelve resolved shear stresses for the zinc-blende structure.

The limitation of this approach is that it ignores the fact that the generation and movement of dislocations in semiconductors is a thermally activated process. Further, the CRSS is a function, not only of the temperature, but also of the strain rate as shown by Müller et al. [207]. Lambropoulos [208] deduced, from simulation studies, that once the crystal length had reached about four times its radius, all the significant stress variation occurred close to the interface and the extent of inelastic deformation was so large that it needed to be coupled into the solution for the thermal stresses near the crystal-melt interface.

A rigorous approach to the problem requires that the temporal variation of the thermal distribution in the crystal is used in the calculation of the thermal stress, creep strain and dislocation density. This is impossibly difficult and simplified models are required. This problem has been addressed by Alexander and Haasen [209] who have developed a phenomenological constituitive equation that describes the motion and multiplication of dislocations in the presence of an applied stress field and relates the formation and interaction of these dislocations to the rate of generation of plastic strain in the material.

Volkl and Müller [210] applied this theory to two indium phosphide crystals grown under different but precisely determined thermal conditions character-ised by a thin and very thick layer of boric oxide respectively. Measured and calculated values for the two crystals are shown in Fig. 10.4. Excellent agree-ment with the model is obtained.

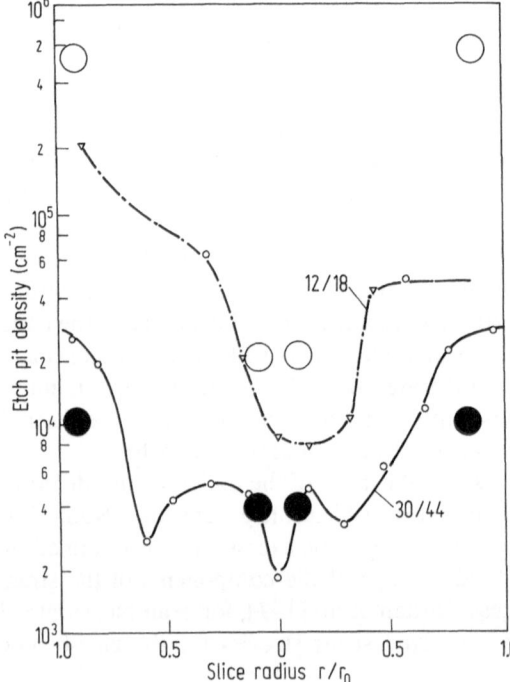

Fig. 10.4. Dislocation density in pulled 30 mm diameter ⟨111⟩ or-iented single crystals of indium phosphide; experimental measure-ments (*large full and open circles*) and model calculations (*curves*). [210]

Maroudas and Brown [219] have carried out an asymptotic analysis of the Alexander and Haasen model appropriate for LEC growth conditions and have shown that dislocation generation is the predominant mechanism for plastic relief of thermally induced stress and that interactions between the dislocations themselves can be largely ignored. The rate of formation of dislocations is much greater than the rate of motion of the crystal through the imposed temperature field. In consequence the dislocation density quickly reaches a value where plastic relief caused by the dislocations balances the applied stress so that dislocation multiplication ceases. The authors show that this limit of fast dislocation formation can be characterised by a dimensionless number which they designate the dislocation Damkohler number (Dk). In the limit of large Dk they obtain asymptotic results for the bulk dislocation density.

The Jordan approach, which ignores the strain rate dependence of the CRSS, implies that thermally-activated rate processes determine the mechanical response of the material. Estimating the dislocation density to be proportional to the excess stress is to assume that the dislocations are produced by shear-induced crystallographic glide and that plastic deformation of the crystal is not an important mechanism for stress relief. The Alexander and Haasen model is an attempt to account for the micromechanical effects of the dislocation density on the plastic stress relief during growth.

At large Dk one can think of there being a short development length, or boundary layer in the crystal adjacent to the crystal-melt interface in which the dislocation density evolves to a value which is in equilibrium with the local stress field. Throughout the bulk of the crystal the structure of the dislocation field can then be estimated from a knowledge of only the magnitude and the gradient of the thermo-elastic stress. For large Dk dislocations are formed rapidly and the dislocation density causes rapid plastic relief which eventually halts the formation of more dislocations after a small length of crystal has been grown.

However, Motakef [261] has argued that the Maroudas and Brown model does not correctly describe the behaviour at low thermal stress where the dislocation density is low. He identifies two limiting cases. In the first, the just-grown layer is treated as isothermal, subject to stresses associated with its accretion onto a stressed elastic body. Since it is isothermal, the stresses must be the result of the force balance in the whole crystal.

In the second limit, the plastic deformation of the crystal is assumed to have effectively removed all stresses at the growth front. Motakef refers to these limiting cases as Stage I and Stage II deformation respectively. His Stage I is similar in essence to the Maroudas and Brown model. Stage II models the evolution of the grown layer as it is subjected to a time-varying thermal strain as the crystal increases its length and cools.

Motakef argues that, in this low dislocation density limit, dislocation multiplication at the growth front is weak and, for this situation, Stage II deformation is the more important. This remains, at present, a controversial issue and one which, because of uncertainties in the relevant physical parameters (in particular the dislocation multiplication coefficient) for GaAs and InP, cannot readily be resolved by recourse to experiment.

Elliott and co-workers [211] have shown that growing GaAs, under conditions which give rise to only a small thermo-elastic stress, in a hot-walled chamber within which a partial pressure of arsenic is equilibrated with the cooling crystal, can yield nearly dislocation-free crystals. The residual dislocations were believed, by the authors, to have resulted from heterogeneous stresses arising at the crystal-melt interface due to gradients of point defect concentration originating from non-uniform incorporation across growth steps on the interface rather than from thermal stresses.

Volkl and Müller [210] also considered the effect of pulling the crystal at high speeds corresponding to a small value of Dk. This gives the dislocations only a very short time to move before the local region of crystal has cooled significantly so that multiplication is again negligible. This is in line with observations made by Dash [218] on the growth of silicon crystals many years ago that dislocation-free ingots can be obtained by using a high pulling speed (in the range 3–10 mm per minute) during the necking procedure following dipping of the seed. He showed that, in addition, already existing dislocations introduced for example from the seed, grow out towards the surface of the crystal at these high pulling rates so that both the propagation of existing dislocations and the generation of new ones are suppressed. Volkl and Müller calculated the critical growth rate needed to achieve this in indium phosphide. Unfortunately for the practical crystal grower, they found that this critical growth rate was several hundred centimetres per hour, way beyond the regime which can be accessed experimentally. Only a reduction in stress or a hardening of the lattice therefore will lead to significant lowering of the dislocation density. Similar conclusions apply to gallium arsenide.

10.3 Point Defect Equilibria in the Cooling Crystal

This subject is a very material-specific topic and we confine ourselves to briefly illustrating the types of phenomena which can occur in cooling Czochralski crystals by taking one specific example; that of gallium arsenide.

The phase diagram of gallium arsenide is frequently shown with a single vertical line representing the solidus implying that it has zero phase extent (Fig. 10.5). One would infer from such a diagram therefore, that a crystal grown at some temperature between the melting point and the arsenic eutectic temperature would have the same composition irrespective of whether it was grown from a gallium-rich or an arsenic-rich solution. In reality, the composition of the crystal will differ slightly; that is to say the solidus has a finite phase extent. Since this phase extent is very small we can consider the crystal as otherwise a perfect crystal lattice but having small concentrations of native point defects. It has been shown experimentally by Bublik et al. [212] that the

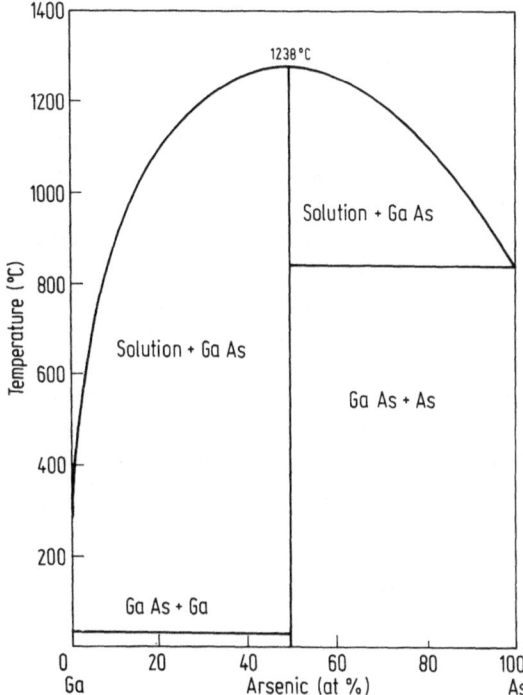

Fig. 10.5. Phase diagram of the Ga-As system. (Rowland [249])

dominant ones grown into the crystal from the melt are arsenic interstitials and arsenic vacancies. The amount by which the crystal deviates from its stoichiometric composition will then be given by the difference in these two point defect concentrations, viz:

$$\delta = [As_i] - [V_{As}] \tag{10.1}$$

The total number of these point defects in equilibrium with the crystal will decrease with decreasing temperature as the crystal cools since their formation by the Frenkel reaction:

$$0 = As_i + V_{As} \tag{10.2}$$

is thermally activated.

The law of mass action, applied to Eq. 10.2 shows that the product of the concentration of interstitials and vacancies is equal to the Frenkel mass action constant (K_{As}):

$$K_{As} = [As_i][V_{As}] \tag{10.3}$$

The equilibrium value of δ will be determined by the activity of arsenic in the ambient phase (the melt). For convenience this is here expressed in terms of a pressure of arsenic dimers which is in equilibrium with that melt, namely:

$$1/2 As_2(g) = As(1) \tag{10.4}$$

The mass action constant for this reaction is:

$$K_{As} = [As_L]p_{As2}^{-1/2}$$ 10.5

where p_{As2} is the partial pressure of the arsenic dimers. From Fig. 10.6, we see that this arsenic-dimer pressure differs by several orders of magnitude between the gallium and arsenic-rich liquidus surfaces at typical epitaxial growth temperatures. The arsenic interstitial concentration will also vary by a large ratio. However, since the product of the interstitial and vacancy concentration is a constant at a given temperature (Eq. 10.3) then the vacancy concentration is reciprocally related to the interstitial concentration. In consequence the value of δ varies strongly as one moves from the arsenic-rich to the gallium-rich liquidus. This is shown schematically in Fig. 10.7. The first point to note is that the congruent point; – that is the composition at which the solidus and liquidus curves touch, is not in general at the stoichiometric composition. In the case of gallium arsenide it is likely that the congruent point is to the gallium-rich side of stoichiometry but some uncertainty on this point remains at present [213]. If we grow a crystal from a melt slightly to the gallium-rich side of congruency, then we see that, as the crystal is cooled to room temperature, the solid solution of excess arsenic vacancies becomes supersaturated. If the temperature at which

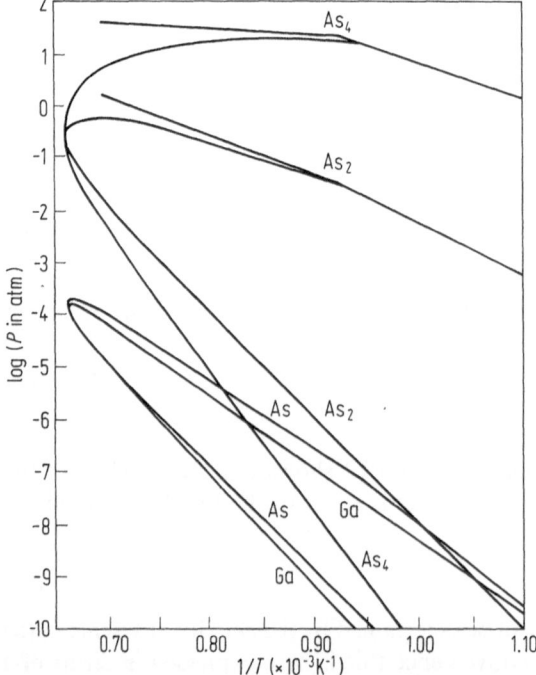

Fig. 10.6. Arsenic dimer and tetramer and gallium monomer partial pressures in equilibrium with the liquidus of the Ga-As system as a function of temperature. (Rowland [249])

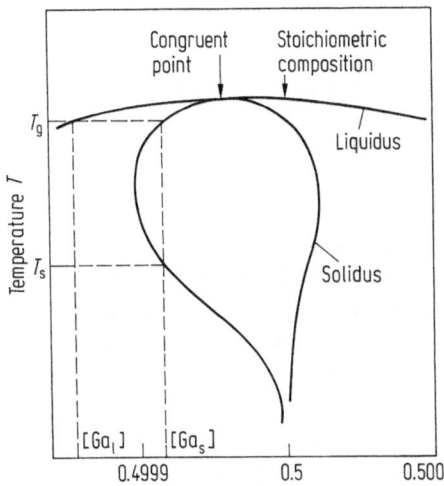

Fig. 10.7. Schematic representation of retrograde solid solubility of Ga in GaAs. The ordinate is the atom fraction of arsenic in the melt. The congruent point is shown to be on the gallium-rich side of stoichiometry. This assignment is tentative

supersaturation first occurs is sufficiently low for the native defects to be immobile within the lattice, then this supersaturation is preserved and the crystal is, at room temperature, in a metastable state. If, however, the defects are sufficiently mobile and the cooling rate sufficiently slow, then precipitation processes can occur so that, at room temperature, the crystal lattice itself is nearly stoichimetric but embedded within it are precipitates associated with the excess component.

Further note that if we directionally solidify a melt whose composition is not congruent, then there is segregation of the excess component into that melt so that progressively the crystal deviates in composition from that corresponding to the congruent point. There will be a boundary layer of the rejected excess constituent ahead of the growing crystal and, if there is turbulence in the melt or some other perturbation, the crystal will contain a striated distribution of crystal composition. This can result in a banded distribution of the precipitated component(s).

The precipitates could in principle nucleate homogeneously but in practice are much more likely to be formed by heterogeneous nucleation on dislocations in the lattice. Moreover, the condensation of elemental native point defects on the dislocations can give rise to dislocation climb, possibly with the emission of other elemental native defects [223]. It is therefore clear that the microstructural properties of the crystal will depend both on its dislocation structure which, as we have seen, in turn depends on the thermal stress during cooling as well as on the rate of cooling which determines the extent to which the crystal can remain close to thermodynamic equilibrium. These points are illustrated well in Fig. 10.1 which shows precipitates of hexagonal arsenic on dislocations in an LEC-grown crystal of gallium arsenide. The dislocations are arranged in a polygonal pattern which is the result of climb of dislocations generated initially on their slip plane systems (Fig. 10.3).

The reaction equations above were written assuming that the point defects were uncharged. In reality most native defects exist in GaAs in one or more charge states, depending on the position of the Fermi level. Coulombic attraction between elemental defects of opposite charge gives rise to defect complex formation during cooling and, since the charge state of the defect depends on the position of the Fermi level, the concentration of such complexes in turn will be Fermi-level and hence dopant-concentration dependent.

Thus far we have considered only defects on the As sub-lattice; less is known about the defects on the Ga sub-lattice. Ga_i are believed to exist with a positive charge but not to be present as stable isolated entities at room temperature. V_{Ga} are believed to be deep acceptors in n-type material but there is no clear evidence for their existence in significant concentrations at room temperature.

One particular native defect known as EL2, whose exact structure is a matter of controversy but about which it is agreed that it contains the arsenic antisite As_{Ga} – i.e. an arsenic atom located on a gallium sub-lattice site – is known to control the semi-insulating behaviour of undoped LEC-grown crystals. It is believed that this defect is formed during cooling of the crystal rather than being grown-in at the melting point [214], [215]. Further, it is known that the concentration of EL2 is approximately a factor of two greater adjacent to the polygonal dislocation array than in the centre of the polygonal structure remote from the dislocation tangles [216]. This suggests that extra antisites somehow arise as a result of the climb of the dislocations, although the details of the mechanism are not at present clear.

We see therefore that the thermodynamics and kinetics of the point defect reactions which occur as a crystal cools can have a profound effect on the crystal properties. Indeed, if we take a semi-insulating gallium arsenide crystal and heat it to 1100 °C followed by rapid cooling to room temperature, the EL2 is seen to have disappeared and the crystal is no longer semi-insulating. Semi-insulating behaviour can be restored by re-annealing the crystal at a somewhat lower temperature (around 900 °C) and then slowly cooling to room temperature [217].

This somewhat superficial review of point defects in gallium arsenide serves to illustrate how control of the process from a commercial standpoint requires control of the cooling phase of the crystal ever bit as much as of the growth phase. Insufficient attention has been paid to this in III/V crystal growth to date. A crucial, largely unanswered, question is what are the critical cooling ranges over which the cooling rate should be controlled? Is slow cooling, to maintain local equilibrium or rapid cooling to quench-in a metastable state required? The answers to these questions depend of course on the uses to which the crystal is to be put. In the LEC technique there is a large discontinuity in cooling rate as the crystal emerges through the encapsulant.

11 Future Developments

What future extensions of the technology can be envisaged? To attempt to answer this, consideration has to be given material by material.

In the case of silicon, size and uniformity are paramount. To this end it is likely that the process will be modified from a strictly batch process to a semi-continuous one where the melt is continuously replenished. This is not as straightforward as it may sound, one serious problem being the build up of impurity in the melt as a result of continuous segregation of residual solutes with segregation coefficient less than unity at the crystallising interface..Additionally, the process has to be stopped periodically to remove finite lengths of grown crystal. The limited life of a silica crucible when in contact with molten silicon sets a practical limit to any pseudo-continuous process with this material. Further developments in the control of oxygen and carbon distribution in silicon are required. Achievments of these goals may involve the use of a cusped magnetic field but that is, at present, not clearly resolved.

Reduction of dislocation density is a critical issue with the III/V compounds and, as indicated in the previous chapter, the point of emergence of the crystal from the encapsulant poses particular problems. Total encapsulation of the crystal has been attempted [263] but is not very compatible with the need to grow long crystals for economic efficiency.

Low thermal gradients exacerbate the problems of diameter control and this is an area where improvements are needed. Several approaches are possible. The load cell as a sensor is admirably rugged in the hostile environment of the pressure puller chamber but, as was shown in Chap. 8, is an ill-conditioned measure of crystal radius. Either this sensor has to be used in a more sophisticated way or it has to be replaced by some other sensor, such as the infra-red imaging used for silicon control. Both approaches are currently being addressed.

The use of multi-variate control instead of the single control variable of heater power, as at present, offers scope for improvement. Multi-variate systems offer the potential for more precise control but are generally less robust in operation. Additional controllable variables are pulling speed and applied magnetic field. Both of these have implications for the micro-uniformity of solute distribution but both carry the advantage that they are rapid acting; i.e. not subject to the thermal delays associated with heater power changes.

Developments in oxide growth are most likely to be focussed on extending the range of materials which can be grown by the pulling technique. This includes extension to yet higher temperatures which, in turn, involves melt containment which avoids contacting other materials. This is a formidable problem and one which leaves little scope for tailoring thermal environments to minimise thermal stress for example. Modelling can have an important role in the design of novel heating and containment configurations.

12 References

1. S Zerfoss, LR Johnson and PH Egli. Discussions of the Faraday Society 5 (1949) 166
2. GK Teal and JB Little. Phys. Rev. 78 (1950) 647
3. GK Teal, IEEE Trans. Electron Devices ED-23 (1976) 621
4. J Czochralski, Z. Physik. Chem. 92 (1918) 219
5. HE Buckley, Crystal Growth (Wiley, New York 1951)
6. CD Brandle in Crystal Growth of Electronic Materials. Ed. E Kaldis (North Holland, Amsterdam 1985) p 101
7. E von Gomperz. Z. Physik 8 (1921) 184
8. H Mark, M Polanyi and E Schmid. Z. Physik 12 (1933) 58
9. Transistor Technology, Vol. 1 Ed. HE Bridgers, JH Scaff and JN Shive. (Van Nostrand, New York, 1958)
10. K Nassau and AM Broyer. J. Appl. Phys. 33 (1962) 3064
11. LG van Uitert, FW Swanekamp and S Preziosi. J. Appl. Phys. 32 (1961) 1176
12. W Bardsley and B Cockayne in Crystal Growth. Proc. ICCG1, Boston 1966. Ed. HS Peiser (Pergamon, Oxford 1967) p 109
13. RN Thomas, HM Hopgood, GW Eldridge, DL Barrett and TT Braggins. Solid State Electron. 24 (1981) 387
14. JB Mullin, RJ Heritage, CH Holliday and BW Straughan. J. Crystal Growth 3/4 (1968) 281
15. R Shuttleworth. Proc. Phys. Soc. (London) A63 (1950) 444
16. W Bardsley, FC Frank, GW Green and DTJ Hurle. J. Crystal Growth 23 (1974) 341
17. JK Kristensen and RMJ Cotterill. Phil. Mag. 36 (1977) 437
18. DTJ Hurle. J. Crystal Growth 63 (1983) 13
19. EA Boucher and TGJ Jones. JCS Faraday I, 76 (1980) 1419
20. J Barrett and RH Heil. Solid State Technology Feb. 1974 p 37
21. DA Hukin in Rare Earths in Modern Science and Technology. Vol. 2. (Plenum, New York 1980) p 25
22. TG Digges, RH Hopkins and RG Seidensticker. J. Crystal Growth 29 (1975) 326.
23. P Burggraaf in Semiconductor International Oct 1984 p 54
24. RW Series. J. Crystal Growth 97 (1989) 92
25. H Hirata and K Hoshikawa. J. Crystal Growth 96 (1989) 747
26. R Gremmelmaier. Z Naturforsch 11a (1956) 511
27. JB Mullin, BW Straughan and WS Brickell. J. Phys. Chem. Solids 26 (1965) 782
28. EPA Metz, RC Miller and R Mazelsky. J. Appl. Phys. 33 (1962) 2016
29. JB Mullin, in III–V Semiconductor Materials and Devices. Ed. RJ Malik. (Elsevier, Amsterdam 1987) p 3
30. J Bass and PE Oliver. J. Crystal Growth 3/4 (1968) 286
31. DTJ Hurle in GaAs Integrated Circuits. Ed. J Mun. (BSP Professional Books, Oxford 1988) pp 1–56
32. VI Alexandrov, VV Osiko, AM Prokorov and VM Tatarintsev. Current Topics in Materials Science Vol. 1. Ed. E Kaldis and HJ Scheel (North Holland, Amsterdam 1976)
33. CD Brandle, DC Miller and JW Nielsen. J. Crystal Growth 12 (1972) 195
34. Crystal Growth Processes. JC Brice. Blackie, Glasgow 1986.
35. B Cockayne. Platinum Metals Review 78 (1973) 86
36. W Schmidt. J. Crystal Growth 33 (1976) 203
37. JC Brice. Philips Tech. Rev. 37 (1977) 250
38. DTJ Hurle. Phil. Mag. 13 (1966) 305
39. D Schwabe and AM Scharmann. J. Crystal Growth 52 (1981) 435
40. JR Carruthers and K Nassau. J. Appl. Phys. 39 (1968) 5205
41. T Munakata and I Tamasawa. J. Crystal Growth 106 (1990) 566

42. JS Turner. Buoyancy effects in Fluids. (Cambridge Univ. Press, Cambridge 1973)
43. R Hide and CW Titman. J. Fluid Mech. 29 (1967) 39
44. T von Karman. Z. Angew Math. Mech. 1 (1921) 244
45. WG Cochran. Proc. Cambridge Phil. Soc. 30 (1934) 365
46. ADW Jones. Prog. Crystal Growth and Charact. 9 (1984) 139
47. JT Stuart. Quart. J. Mech. Appl. Math. 7 (1954) 446
48. J Barthel and K Eichler. Kristall und Technik. 2 (1967) 205
49. MH Rogers and GN Lance, J. Fluid Mech. 7 (1960) 617
50. N Riley. Quart. J. Mech. App. Math. 15 (1962) 459
51. ADW Jones. J. Crystal Growth 61 (1983) 235
52. A Bottaro and A Zebib. J. Crystal Growth 97 (1989) 50
53. S Chandrasekhar, Hydrodynamic and Hydromagnetic Stability (Clarendon Press, Oxford, 1961)
54. WE Langlois, Physico Chemical Hydrodynam. 2 (1981) 245
55. RA Brown, A I Ch. E J. 34 (1988) 881
56. M Mihelcic and K Wingerath. J. Crystal Growth 97 (1989) 42
57. DC Miller. in Materials Processing in the reduced gravity environment of space. Ed. GE Rindone (Elsevier, Amsterdam 1982) p 373
58. DC Miller and TL Pernell. J. Crystal Growth 57 (1982) 253
59. ADW Jones. J. Crystal Growth 61 (1983) 70
60. ADW Jones. Phys. Fluids 28 (1985) 31
61. PAC Whiffin, TM Bruton and JC Brice. J. Crystal Growth 32 (1976) 205
62. K Tagaki, T Fukazawa and M Ishii. J. Crystal Growth 32 (1976) 89
63. CD Brandle. J. Crystal Growth 42 (1977) 400
64. B Cockayne, B Lent and JM Roslington. J. Mater. Sci. 11 (1976) 259
65. JR Carruthers. J. Crystal Growth 36 (1976) 212
66. V Nikolov, K Iliev and P Peshev. J. Crystal Growth 89 (1988) 313, 324
67. D Schwabe and J. Metzger. J. Crystal Growth 97 (1989) 23
68. RA Brown, TA Kinney, PA Sackinger and DE Bornside. J. Crystal Growth 97 (1989) 99
69. G Müller and G Neumann. J. Crystal Growth 59 (1982) 548
70. LD Landau and EM Lifshitz. Fluid Mechanics Vol. III (Pergamon, Oxford 1959)
71. D Ruelle and F Takens. Commun. Math. Phys. 20 (1971) 162
72. JMT Thompson and HB Stewart, Non-Linear Dynamics and Chaos. (J Wiley New York 1986)
73. EN Lorenz. J. Atmospheric Sci. 20 (1963) 130
74. HL Swinney. Physica D7 (1983) 3
75. EM Swiggard, SH Lee and FW von Batchelder. Inst. Phys. Conf. Series 33b (1977) 23
76. EM Porbansky. J. Appl. Phys. 30 (1959) 1455
77. B Cockayne. J. Crystal Growth 42 (1977) 413
78. JS Haggerty and JF Wenkus. US Patent No. 3704093 (Nov. 28, 1972)
79. JR Ockendon and AB Tayler, Free Boundary problems: theory and applications. Vol III Eds. A Bossavit, A Damlamian and M Fremond. (Pitman Press, Boston 1985)
80. JC Brice. J. Crystal Growth 2. (1968) 395
81. CM Elliott and JR Ockendon. Weak and Variational Methods for moving boundary Problems. Research Notes in Maths. 59 (Pitman Adv. Publ. Prog., Boston 1982)
82. N Kobayashi, in Preparation and Properties of Solid State Materials. Ed. WR Wilcox (Dekker, New York 1981)
83. EJ Stern. IMA Appl. Math. 35 (1985) 205
84. LJ Atherton, JJ Derby and RA Brown. J. Crystal Growth 84 (1987) 57
85. VA Borodin, LB Davidora, VN Erofeev, AV Shdanov, SA Startsev and VA Tatarchenko J. Crystal Growth 46 (1979) 757
86. TW Hicks. J. Crystal Growth 84 (1987) 598
87. JJ Derby and RA Brown. J. Crystal Growth 76 (1986) 339
88. AS Jordan. J. Crystal Growth 49 (1980) 631
89. AB Crowley. IMA Appl. Math. 30 (1983) 173
90. AB Crowley, EJ Stern and DTJ Hurle. J. Crystal Growth 97 (1989) 697
91. F Dupret, Y Ryckmans P Wouters and MJ. Crochet. J. Crystal Growth 79 (1986) 84
92. AG Ostrogorsky, KH Yao and AF Witt, J. Crystal Growth 84 (1987) 460
93. K Bottcher, A Kruger and B Schleusener. Cryst. Res. Technol. 25 (1990) 1007
94. KF Hulme and JB Mullin. Solid State Electron. 5 (1962) 211
95. JB Mullin and KF Hulme. J. Phys. Chem. Solids 17 (1960) 1

96. B Cockayne. J. Crystal Growth 42 (1977) 413
97. JA Burton, RC Prim and WP Slichter. J. Chem. Phys. 21 (1953) 1987
98. LO Wilson. J. Crystal Growth 44 (1978) 247
99. AA Wheeler. J. Eng. Math. 14 (1980) 161
100. AF Witt, M Lichtensteiger and HC Gatos. J. Electrochem. Soc. 120 (1973) 1119
101. DTJ Hurle and E Jakeman. J. Crystal Growth 5 (1969) 227
102. DTJ Hurle, E Jakeman and ER Pike. J. Crystal Growth 3, 4 (1968) 633
103. BI Birman. Growth of Crystals. Vol. 11 Ed. AA Chernov Consultants Bureau p 278
104. AMJG van Run. J. Crystal Growth, 54 (1981) 195
105. NA Avdonin, EN Martazan, DG Ratnokov and GA Goryushin. Izv. Akad. Nauk SSSR, Neorg. Mater. 13 (1977) 941
106. LO Wilson. J. Crystal Growth 48 (1980) 435, 451
107. AA Wheeler. J. Crystal Growth 56 (1982) 67
108. PAC Whiffin and JC Brice. J. Crystal Growth 10 (1971) 91
109. D Mateika, in Advanced Crystal Growth Eds. PM Dryburgh, B Cockayne and KG Barraclough. (Prentice Hall, New York 1987) p 149
110. JP Wallace, JK Tien, JA Stefani and KS Choe. J. Appl. Phys. 69 (1991) 550
111. RS Wagner. Acta. Met. 8 (1960) 57
112. HP Utech and MC Flemings. J. Appl. Phys. 37 (1966) 2021
113. HA Chedzey and DTJ Hurle. Nature. 210 (1966) 933
114. K Hoshikawa, H Kohda, H Hirata and H Nakanishi. Jap. J. Appl. Phys. 19 (1980) L33
115. T Suzuki, N Isawa, Y Okubo and K Hoshi. Proc. 4th Int. Symp. on Silicon, Materials Science & Technology: Semiconductor Silicon 1981 (ECS Pennington, NJ 1981)
116. RW Series and DTJ Hurle. J. Crystal Growth, 113 (1991) 305
117. S Makram-Ebeid, P Langlade and GM Martin in Semi-insulating III–V Materials. Ed. DC. Look and JS Blakemore. (Shiva Publishing Ltd., Nantwich, UK 1984) p 184
118. AF Witt, CJ Herman and HC Gatos. J. Mater. Sci. 5 (1970) 822
119. KM Kim and P Smetana. J. Appl. Phys. 58 (1985) 2731
120. L-N Hjellming and JS Walker. J. Fluid Mech. 182 (1987) 335
121. L-N Hjellming. J. Crystal Growth 104 (1990) 327
122. L-N Hjellming and JS Walker. J. Crystal Growth 92 (1988) 371
123. WE Langlois and K-J Lee. J. Crystal Growth 62 (1983) 481
124. LN Hjellming and JS Walker. J. Crystal Growth 164 (1986) 237
125. WE Langlois. J. Crystal Growth 83 (1987) 51
126. RN Thomas, HM Hopgood, PS Ravishankar and TT Braggins. J. Crystal Growth 99 (1990) 643
127. T Kakutani. J. Phys. Soc. Japan 17 (1962) 1496
128. DTJ Hurle and RW Series. J. Crystal Growth 73 (1985) 1
129. RA Cartwright, DTJ Hurle, RW Series and J Szekely. J. Crystal Growth 82 (1987) 327
130. S Kobayashi. J. Crystal Growth 75 (1986) 301
131. MG Williams, JS Walker and WE Langlois. J. Crystal Growth 100 (1990) 233
132. PS Ravishankar, TT Braggins and RN Thomas. J. Crystal Growth 104 (1990) 617
133. GA Rozgonyi and CW Pearce, Appl. Phys. Lett. 31 (1977) 343
134. KG Barraclough. EMIS Datareviews No.4. Properties of Silicon (INSPEC, London 1988) p 291
135. R Series. J. Crystal Growth 97 (1989) 85
136. H Hirata and K Hoshikawa. J. Crystal Growth 106 (1990) 657
137. DJ Barker. J. Appl. Phys. 35 (1964) 398.
138. K Terashima and T Fukuda. J. Crystal Growth 63 (1983) 423
139. K Hoshikawa, H Kohda and H Hirata. Japan J. Appl. Phys. 23 (1984) L37
140. J Osaka, H Kohda, T Kobayashi and K Hoshikawa. J. Appl. Phys. 23 (1984) L195
141. H Kodha, K Yamada, H Nakanishi, T Kobayashi. J. Osaka and K Hoshikawa. J. Crystal Growth 71 (1985) 813
142. D Hofmann, M Mosel and G Muller. Semi-insulating III–V Materials Malmo, Sweden 1988 Ed. G Grossmann and L Ledebo (IOP Publishing, Bristol 1988) p 429
143. S Ozawa, T Kimura, J Kobayashi and T Fukuda. Appl. Phys. Lett. 50 (1987) 329
144. D Hofmann, D Mosel and G Muller, in Semi-insulating III–V Materials, Malmo, Sweden 1988, Ed. G Grossmann and L Ledebo (IOP Publishing, Bristol 1988) p 429
145. S Kobayashi and T Mugurama. J. Crystal Growth 84 (1987) 559
146. JB Mullin and KF Hulme. J. Electronics and Control 4 (1958) 170
147. T Surek. J. Appl. Phys. 47 (1976) 4384

148. T Surek and SR Coriell. J. Crystal Growth 37 (1977) 253
149. VA Tatarchenko. J. Crystal Growth 37 (1977) 272
150. DTJ Hurle, GC Joyce, M Ghassempoory, AB Crowley and EJ Stern. J. Crystal Growth 100 (1990) 11
151. DTJ Hurle, GC Joyce, GC Wilson, M Ghassempoory and C Morgan. J. Crystal Growth 74 (1986) 480
152. S Vojdami, AE Dabiri and H Ashoori. J. Crystal Growth 24/25 (1974) 374
153. EJ Patzner, RG Dessauer and MR Poponiak. SCP and Solid State Tech. Oct. 1967, p 201
154. U Gross and R Kersten. J. Crystal Growth 15 (1972) 85
155. KJ Gartner, KF Rittinghaus and A Seeger. J. Crystal Growth 13/14 (1972) 619
156. DF O'Kane, TW Kwap, L Gulitz and AL Bednowitz. J. Crystal Growth 13/14 (1972) 624
157. HD Pruett and HY. Lien. J Electrochem. Soc. 121 (1974) 822
158. HJA van Dijk, CMG Jochern, GJ Scholl and P van der Werf. J. Crystal Growth 21 (1974) 310
159. W Bardsley, GW Green, CH Holliday and DTJ Hurle. J. Crystal Growth 16 (1972) 277
160. DTJ Hurle. J. Crystal Growth 42 (1977) 473
161. W Bardsley, GC Joyce and DTJ Hurle. J. Crystal Growth 40 (1977) 13
162. W Bardsley, DTJ Hurle, GC Joyce and GC Wilson. J. Crystal Growth 40 (1977) 21
163. AV Stepanov, Sov. Phys. Techn. Phys. 4 (1959) 339
164. PI Antonov. J. Crystal Growth 23 (1974) 318
165. EP Dubnik, DI Levizon, AG Petrik and VV Selin. Bull. Acad. Sci. USSR, Phys. Ser. 37 (1973) 27
166. Yu. M Smirnov, Bull. Acad Sci. USSR Phys. Ser. 33 (1969) 1825
167. LP Egorov, LS Okun, LM Zatulovskii, PM Chaikin, VV Gulaev, V Yu Zhvirtlyanskii, DI Levinzon, Yu M Smirnov and GV Sachkov. Bull. Acad. Sci. USSR, Phys. Ser. 37 (1973) 12
168. VA Tatarchenko. J. Crystal Growth 37 (1977) 272
169. HE LaBelle, Jr. and AI Mlavsky, Mater Res. Bull. 6 (1971) 571, 581.
170. M Cole, RM Ware and MA Whittaker Proc. ECCG1, Zurich 1976 Crystal Growth and Materials Ed. E Kaldis and HJ Scheel (North Holland, Amsterdam 1977)
171. GC Joyce, DTJ Hurle and QAE Vaughan. J. Crystal Growth 132 (1993) 1
172. W Bardsley, GW Green, CH Holliday, DTJ Hurle, GC Joyce, WR MacEwan and PJ Tufton. Inst. Phys. Conf Series No. 24 (1975) 355
173. D Rumsby, Metals Research Semiconductors. Private communication.
174. W Bardsley, JM Callan, HA Chedzey and DTJ Hurle. Solid State Electronics 3 (1961) 142
175. JW Rutter and B Chalmers, Canad. J. Phys. 31 (1953) 15
176. WA Tiller, KA Jackson, JW Rutter and B Chalmers. Acta Met. 1 (1953) 428
177. DTJ Hurle, Progress in Materials Science Vol. 10 (Pergamon Press, Oxford 1962) p 79
178. WG Pfann, Zone Melting (John Wiley, New York 1966) p 254
179. WW Mullins and RF Sekerka. J. Appl. Phys. 35 (1964) 444
180. ME Glicksman, SR Coriell and GB McFadden. Annual Review of Fluid Mechanics 18 (1986) 307
181. SH Davis. J. Fluid Mech. 212 (1990) 241
182. BI Birman, in Growth of Crystals Vol. 7. Ed. NN Sheftal (Consultants Bureau, New York 1969)
183. SR Coriell, DTJ Hurle and RF Sekerka. J. Crystal Growth 32 (1976) 1.
184. M Hennenberg, A Rouzaud, D Camel and JJ Favier. J. Crystal Growth 85 (1987) 49
185. SR Coriell, GB McFadden, RF Boisvert and RF Sekerka. J. Crystal Growth 69 (1984) 15
186. RT Delves. J. Crystal Growth 8 (1971) 13
187. K Brattkus and SH Davis. J. Crystal Growth 87 (1988) 385
188. R Singh, AF Witt and HC Gatos. J. Electrochem. Soc. 121 (1974) 380
189. AA Wheeler. J. Crystal Growth 67 (1984) 8
190. M Duseaux and G Jacob. Appl. Phys. Lett. 40 (1982) 790
191. LG Eidelman. Crystal Res. and Technol. 18 (1983) 405.
192. DAO Hope, MS Skolnick, B Cockayne, J Woodhead and RC Newman. J. Crystal Growth 71 (1985) 795
193. S McGuigan, RN Thomas, DL Barrett, GW Eldridge, RL Messham and BW Swanson. J. Crystal Growth 76 (1986) 217
194. MG Tabache, ED Bourret and AG Elliot. Appl. Phys. Lett. 49 (1986) 289
195. AS Jordan, AR von Neida and R Caruso. J. Crystal Growth 70 (1984) 555
196. MG Mil'vidskii and BI Golovin. Sov. Phys. Solid State. 3 (1961) 737
197. AS Jordan, R Caruso and AR von Neida. Bell System Tech. J. 59 (1980) 593
198. S Motakef and AF Witt. J. Crystal Growth 80 (1987) 37

199. GO Medouye, DJ Bacon and KE Evans. J. Crystal Growth 88 (1988) 397
200. GO Medouye, KE Evans and DJ Bacon. J. Crystal Growth 97 (1989) 709
201. GO Medouye, DJ Bacon and KE Evans. J. Crystal Growth 108 (1991) 627
202. S Motakef. J. Crystal Growth 96 (1989) 201
203. C Schrizov, IV Samarasekera and F Weinberg. J. Crystal Growth 85 (1987) 142
204. M Duseaux. J. Crystal Growth 61 (1983) 576
205. G Szabo. J. Crystal Growth 73 (1985) 131
206. N Kobayashi and T Iwaki. J. Crystal Growth 73 (1985) 96
207. G Muller, R Rupp, J Volkl, H Wolf and W Blum. J. Crystal Growth 71 (1985) 771
208. JC Lambropoulos. J. Crystal Growth 80 (1987) 245
209. H Alexander and P Haasen, in: Solid State Physics Vol. 21. Eds. F Seitz, D Turnbull and H Ehrenreich (Academic Press, New York 1968) p 27
210. J Volkl and G Müller. J. Crystal Growth 97 (1989) 136
211. AG Elliot, D Vanderwater and C-L Wei, Mater. Sci. Eng. B1 (1988) 23
212. VT Bublik, VV Karataev, RS Kulagin, MG Mil'vidskii, VB Osvenskii, OG Stolyarov and LP Khnolodnyi. Sov. Phys. Crystallog. 18 (1973) 218
213. DTJ Hurle, 5th Conf. on Semi-insulating III–V Materials, Malmo 1988 Ed. G Grossmann and L Ledebo (Adam Hilger, Bristol 1988) p 11
214. J Lagowski, HC Gatos, JM Parsey, K Wada, M Kaminsko and W Walukiewicz. Appl. Phys. Lett. 40 (1983) 342
215. DTJ Hurle, Semi-insulating III–V Materials Malmo 1988. Ed. G Grossmann and L Ledebo (Adam Hilger, Bristol 1988) p 11
216. MR Brozel, I Grant, RM Ware, DJ Stirland and MS Skolnick. J. Appl. Phys. 56 (1984) 1109
217. S Clark, MR Brozel, DJ Stirland, DTJ Hurle and I Grant. Defects in Semiconductors Japan 1989
218. WC Dash, Growth and Perfection of Crystals. Proc. Int. Conf. on Crystal Growth, Co-operstown (1958) (J Wiley, NY 1958) 361
219. D Maroudas and RA Brown. J. Crystal Growth 108 (1991) 399
220. P Petroff and RL Hartman. J Appl. Phys. 45 (1974) 3899
221. Y Matsouka, K Ohwada and M Hiriyama, IEEE Trans on Elect. Devices. ED-31 (1984) 1062
222. B Cockayne, DS Robertson and W Bardsley. Brit. J Appl. Phys. 15 (1964) 1165
223. PM Petroff and LC Kimerling. Appl. Phys. Lett 29 (1976) 461
224. D Rumsby and S Lawton, private communication
225. RC Dorf. Modern Control Systems (Addison- Wesley, Reading Mass. 1976) Ch 10
226. E Zalewski and J Zmija. Acta. Physica. Polonica. A51 (1977) 807, 819
227. J Kvapil, J Kubelka, J Kvapil and B Perner. Cryst. Res. Technol. 18 (1983) 695
228. WF Leverton. J Appl. Phys. 29 (1958) 1241
229. J Goorissen Phillips Tech. Rev. 21 (1960) 185
230. JC Brice. Selected Topics in Solid State Physics Vol. 5. The Growth of Crystals from the Melt (North Holland, Amsterdam 1965) p 150
231. K Terashima, H Nakajima and T Fukuda. Japan. J Appl. Phys. 21 (1982) L452
232. GA Satunkin and AG Leonov. J. Crystal Growth 102 (1990) 592
233. KM Kim, A Kran, P Smetana and GH Schwuttke. J. Electrochem. Soc. 130 (1983) 1156
234. K Riedling. J. Crystal Growth 89 (1988) 435
235. MA Gevelber, MJ Wargo and G Stephanopoulos. J. Crystal Growth 85 (1987) 256
236. H Schlichting, Boundary Layer Theory. (McGraw Hill NY 1968)
237. B Cockayne, M Chesswas and DB Gasson. J. Materials Sci. 4 (1969) 450
238. JAM Dikhoff. Solid State Electronics. 1 (1960) 202
239. B Cockayne and MP Gates. J Materials Sci. 2 (1967) 118
240. DTJ Hurle and B Cockayne. in Characterisation of Crystal Growth Defects by X-ray Methods. Ed. BK Tanner and DK Bowen. NATO Advanced Study Institute Series B Vol. 63. (Plenum Press, New York 1980) p 46
241. AJ Goss, KE Benson and WG Pfann. Acta Met. 4 (1956) 332
242. HA Labelle and AI Mlavsky. Mater. Res. Bulletin 6 (1971) 571, 581.
243. DTJ Hurle, O Jones and JB Mullin. Solid State Electronics. 3 (1961) 317
244. W Bardsley, DTJ Hurle, M Hart and AR Lang. J. Crystal Growth 49 (1980) 612
245. W Bardsley, B Cockayne, GW Green and DTJ Hurle. Solid State Electronics. 61 (1963) 389
246. M Hennenberg, A Rouzaud, D Camel and J-J Favier. J. Physique 48 (1987) 173
247. GJ Merchant and SH Davis. J. Crystal Growth 96 (1989) 737
248. AG Cullis, PD Augustus and DJ Stirland.J. Appl. Phys. 51 (1980) 2556

249. MC Rowland in Semiconductor Lasers Ed. C Gooch (Wiley Interscience, New York 1989) p 133
250. Semiconductors and Semimetals 20 Ed. RK Willardson and AC Beer. (Academic Press, New York 1984) p 88.
251. B Cockayne and JM Roslington. J.Materials Sci. 8 (1973) 601
252. A Muhlbauer, R Kappelmeyer and F Keiner. Z Naturforsch. 20a (1965) 1089
253. A Muhlbauer. Z Naturforsch. 21a (1966) 490
254. AA Wheeler. Proc. Roy. Soc. A379 (1982) 327
255. A Murgai, HC Gatos and WA Westdorp. J. Electrochem. Soc. 126 (1979) 2240
256. G Müller, G Neumann and W Weber. J. Crystal Growth 119 (1992) 8
257. N Riley and DA Sweet. J. Crystal Growth 47 (1979) 557
258. DJ Stirland. private communication.
259. J Baumgartl and G Müller. Proc. 8th European Symposium on Materials and Fluid Sciences in Microgravity, April 1992. ESA Publication SP333 (1993) p 161
260. W Uelhoff. J. Crystal Growth 65 (1983) 278
261. S Motakef. J. Crystal Growth, 114 (1991) 47
262. G Müller. Crystal growth from the melt. (Springer-Verlag, Berlin 1988) p 1
263. T Ibuka, Y Seta, M Tanamura, F Orito, T Okano, G Hyuga and J Osaka in Semi-insulating III–V Materials, Hakone 1986. Ed. H Kukimoto and S Miyazawa. (OHM-North Holland, Amsterdam 1986) p 77
264. K Kakimoto, M Watanabe, M Eguchi and T Hibiya. J. Crystal Growth 126 (1993) 435
265. R Hide and PJ Mason, Advan. Physics 24 (1975) 47
266. ET Eady. Tellus 1 (1949) 33
267. RB Lambert and HA Snyder. J. Geophys. Res. 71 (1966) 5225
268. G Buzyna, RL Pfeffer and R Kung. J. Fluid Mech. 145 (1984) 377

Subject Index

adaptive control, 100
after-heaters, 126, 129
alumina, 20
ambient gas turbulence, 99
anomalous materials, 100
arsenic interstitials, 133
arsenic precipitates, 126
arsenic reservoir, 17
arsenic vacancies, 133, 134
automatic diameter control, 63, 86
axial field, 66, 76
axial magnetic field, 68, 74, 101
azimuthal motion, 68

baffles, 61
baroclinic instability, 32
baroclinic waves, 32
Biot number, 40
bismuth germanate, 88
bismuth silicon oxide, 21, 34, 36
boric oxide, 15, 17, 19, 45, 76, 128–130
boundary layer thickness, 53
boundary layer, 49, 75
– uniformly accessible, 49,
Boussinesq approximation, 33
Brewster angle, 47
bright ring technique, 98
bright ring, 13, 88
bulk flow, 69
buoyancy driven convection, 21, 23, 74
buoyancy-driven flow, 33
Burger's vector, 125
Burton, Prim and Slichter, 52
– relationship, 70
– equation, 104

calcium fluoride, 39
calcium tungstate, 3, 39, 113, 114, 117
capillary constant, 120
capillary effects, 79
cell boundaries, 109, 111
cellular interface, 125
cellular morphology, 110
cellular structure, 105, 106, 108, 110, 117, 118
centrifugal force, 26
Chalmers number, 122
chaos, 38
– route to, 38
chaotic fluctuations, 76

Cochran flow, 69
Cochran solution, 69
cold crucible, 11
commercial high pressure LEC puller, 18
commercial pullers, 11
configured field, 66
congruent point, 134, 135
constitutional supercooling, 21, 101, 103, 104,
        107, 109, 113, 121, 123
contacting angle, 78
control signal, 98
control strategies, 99
convective turbulence, 38
coracle, 92
Coriolis force, 26, 27, 38
corrugations, 106, 107
critical resolved shear stress, 60, 128
critical wave number, 121
cross-correlation, 98
crucible materials, 11
crucible rotation, 10, 77
crucible weighing, 90, 93, 99
crucible, 16, 44, 55, 61, 68, 74, 87, 90
– inner floating, 55
– inner, 61
– wall temperature, 68
crucibles, 20, 21
– precious metal, 20, 21
crystal imaging, 88
crystal facet, 51
crystal pulling, 78
crystal rotation rate, 71
crystal rotation, 10, 13, 39, 68, 101,
crystal shape, 79
crystal shoulder, 98
crystal stoichiometry, 63, 75
cusped magnetic field, 14, 67, 77, 137
cylindrical crystal, 78
Czochralski crystal, 40
Czochralski growth, 25, 31, 32, 42, 43, 52, 56, 79,
        84, 93, 122
Czochralski melt, 33, 37–39, 64, 67, 104
Czochralski process, 86
Czochralski technique, 22, 61
Czochralski's apparatus, 1
Czochralski, 2

Damkohler number, 131
damping force, 65

148   Subject Index